Yuan Jiang

Solving the Inverse Problem of Electrocardiography in a Realistic Environment

Vol. 9
Karlsruhe Transactions on Biomedical Engineering

Editor:
Karlsruhe Institute of Technology
Institute of Biomedical Engineering

Solving the Inverse Problem of Electrocardiography in a Realistic Environment

by
Yuan Jiang

Dissertation, Karlsruher Institut für Technologie
Fakultät für Elektrotechnik und Informationstechnik, 2010

Impressum

Karlsruher Institut für Technologie (KIT)
KIT Scientific Publishing
Straße am Forum 2
D-76131 Karlsruhe
www.uvka.de

KIT – Universität des Landes Baden-Württemberg und nationales
Forschungszentrum in der Helmholtz-Gemeinschaft

KIT Scientific Publishing 2010
Print on Demand

ISSN: 1864-5933
ISBN: 978-3-86644-486-7

Solving the Inverse Problem of Electrocardiography in a Realistic Environment

Zur Erlangung des akademischen Grades eines

DOKTOR-INGENIEURS

von der Fakultät für

Elektrotechnik und Informationstechnik

des Karlsruher Instituts für Technologie (KIT)

vorgelegte

DISSERTATION

von

M. Sc. Yuan Jiang
aus Wuhan
China

Tag der mündlichen Prüfung: 11. Februar 2010
Hauptreferent: Prof. Dr. rer. nat. Olaf Dössel
Korreferent: Prof. Dr. Bjørn Fredrik Nielsen

Contents

List of Tables

1

Introduction

1.1 Motivation

Heart disease is the leading cause of death. In the clinical practice a major tool to diagnose and monitor heart diseases is electrocardiogram (ECG). ECG records the difference of electrical potentials on the body surface invoked by the cardiac activities. Thus, it remotely reflects the functionality of the heart. Although the conventional ECG provides important information about the heart status, the interpretation and analysis of ECG require experienced cardiologists.

In order to understand cardiac activities from another perspective, many efforts have been made in the research field of cardiac modeling. The electrophysiology of the heart is studied by experiment at the microscopic level and the electrical and chemical behavior of cardiac cells is analyzed and modeled using mathematical equation tools. Based on the cardiac models the electrical functionality of the heart can be simulated. Furthermore, the corresponding body surface potentials are computed by solving the forward problem of ECG. Through the comparison between the simulated and experimental ECGs the correctness of a cardiac model can be evaluated. The knowledge gained from the evaluation can be used in the further improvement of the cardiac models. In the current thesis a cellular automaton approach is utilized to simulate the bioelectrical activities in the heart. The bidomain model and finite element method are used to solve the forward problem of ECG.

The ability of modeling of cardiac activities and of further calculation of ECG enables a promising noninvasive technique to assess the heart functionality: the inverse problem of electrocardiography. It reconstructs the bioelectrical sources in the heart from the multichannel ECG measured on the body surface. In the solution of the inverse problem the anatomy of the patient's thorax and physical properties of different tissues are also taken into account. The reconstruction of cardiac sources gives an insight into the heart and provides the straightforward information about the heart status. However, the inherent underdetermination and ill-posedness of the inverse problem of ECG limit the quality and stability of the inverse solution. Small errors in the measurement and in the model can lead to unexpected fluctuations in the reconstruction. Therefore, a regularization technique must be deployed to stabilize the inverse solution by introducing additional constraints. In the present thesis the inverse problem of ECG is solved using different regularization methods in a realistic environment instead of in an idealistic environment in order to investigate the influence of different kinds of error on the inverse solution. Among the applied regularization methods, a regularization method combining the spatio-temporal framework and the hybrid framework is proposed and achieves the best result in the realistic environment. Moreover, maximum *a posteriori* (MAP) based regularization is employed and further developed to incorporate simulation results as *a priori* information and to apply the spatio-temporal framework. The MAP-based regularization and its spatio-temporal version are applied to two heart diseases: premature ventricular contraction (PVC) and myocardial infarction.

The second focus of the present thesis is on the application of ECG simulation results in the optimization of ECG measurement systems for the detection and localization of myocardial infarction. The first system under consideration is the multichannel ECG for the inverse problem of ECG. The second one is a wearable ECG monitoring system with a limited number of electrodes to effectively detect ischemic events and myocardial infarction.

Besides the noninvasive approaches, intracardiac measurement is a minimal invasive approach to obtain information about the heart. It allows the registration of an electrogram directly on the heart inner surface using the catheter, which is normally inserted into the heart chamber through a large vein. One study included in the current thesis is the use of computer simulation to support the development of an impedance-based positioning system for the simultaneous localization of multiple catheter electrodes during the interventional procedure.

1.2 Research Topics

The present thesis covers three topics related to the computer simulation of cardiac activities and of the corresponding electrical signals on the body surface:

- The inverse problem of electrocardiography.
- Optimization of electrode positions for the detection and localization of myocardial ischemia and infarction.
- An impedance-based catheter positioning system for cardiac mapping and navigation.

1.3 Structure of the Thesis

Part I (Chapter 2) introduces the fundamental knowledge needed for the understanding of the topics discussed in the current thesis. Included are the anatomy of the human heart, the electrophysiology of cardiac cells, ECG measurement systems and the notation of ECG.

Part II is devoted to the methodology applied in this work:

- Chapter 3 describes the anatomical models used in the investigation, the simulation of cardiac activities using the cellular automaton and the modeling of cardiac diseases.
- Chapter 4 discusses the forward problem of ECG, a realistic environment of the ECG simulation and the optimization of electrode positions for a wearable ECG system using the ECG simulation results.
- Chapter 5 covers the definition of the inverse problem of ECG, the source models, the regularization methods required by the solution of the inverse problem of ECG, and the optimization of electrode positions for a multichannel ECG system for solving the inverse problem.
- Chapter 6 provides a short introduction into the intracardiac mapping and navigation and describes the concept of the novel impedance based catheter positioning system.

Part III presents the results obtained in the course of this thesis:

- Chapter 7 contains the results related to the cardiac modeling and the forward simuation.
- Chapter 8 displays the inverse results obtained using different regularization methods and for different purposes.
- Chapter 9 shows the localization results of the proposed catheter positioning system in the simulation study.

At the end, an outlook for the future work is given in Chapter 10.

Part I

Fundamental Knowledge

Medical Foundations

2.1 Anatomy of the Heart

The heart is one of the most vital organs in the human body. It contracts rhythmically and repeatedly to pump blood through the circulatory system to the entire body. A normal heart is situated underneath the sternum and slightly to the left of the thorax. It weights $250 - 350\ g$ on average and has a size of about an adult's fist. The anatomy of the heart is illustrated in Fig. 2.1. The heart is mainly composed of muscle, which is also called myocardium. The heart is divided into four chambers: two atria and two ventricles. The chambers are filled with blood. Valves between the atria and ventricles as well as between the chambers and blood vessels ensure that the blood flows in a unique direction. The deoxygenated venous blood returns from the systemic circulation through the superior vena cava and the inferior vena cava into the right atrium. Then, the deoxygenated blood enters the right ventricle. Later, it will be pumped into the pulmonary circulation. In the lungs the blood gets refreshed and the oxygenated blood passes through the pulmonary veins into the left atrium after the pulmonary circulation. Afterwards, the blood flows into the left ventricle. With the contraction of the left ventricle the oxygenated blood is expelled into the systemic circulation through the aorta and transferred to the entire body at the end. Because the left ventricle requires much power to pump blood into the systemic circulation, where the pressure is remarkably higher than that in the pulmonary circulation, the left ventricular wall is considerably thicker than the right ventricular wall.

2.2 Electrophysiology of the Heart

2.2.1 Action Potential of Cardiac Cells

A cardiac cell is capable of generating an action potential, i.e., electrical activation across its membrane, on which the electrical excitation of the heart is based. As can be seen in Fig. 2.2 different kinds of cardiac cell have various shapes of action potential. The cells of the sinoatrial node (SA node) and of the components in the excitation conduction system are able to activate spontaneously, because they have non-constant resting potentials. Taking the SA node as example, its resting potential incessantly increases from the minimum at about $-70\ mV$ in the diastolic phase to about $-40\ mV$, which is the threshold potential for the activation. When the threshold potential is reached, an action potential is triggered immediately. The non-constant resting potential is caused by the changes in ion conduction and ionic flows through the cell membrane. After the beginning of the repolarization of the cells in the SA node the non-selective ion conduction

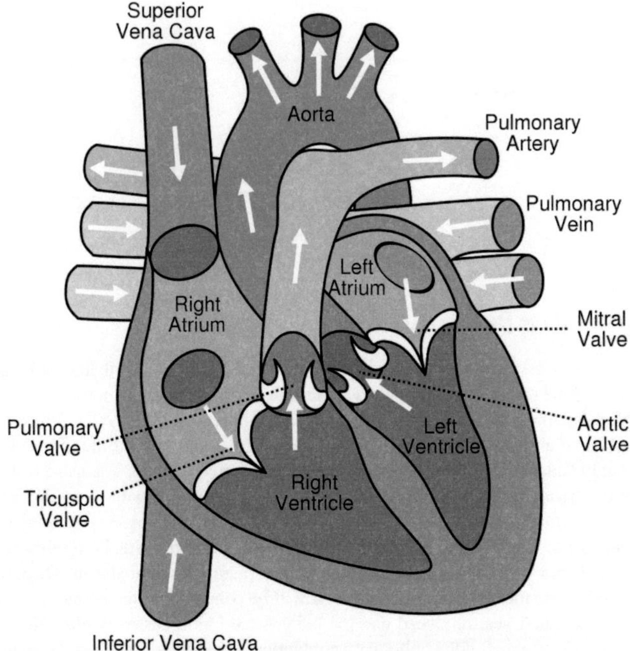

Fig. 2.1. Diagram of the human heart [1].

and the cation flux into the intracellular space begin to grow, which leads to a slow depolarization. Once the depolarization occurs, the conduction for Ca^{2+} increases quickly. It speeds up the influx of Ca^{2+} and results in a rapid increment of the action potential. When the action potential becomes positive, the K^+-channels are abruptly opened and K^+ ions flow outwards. Accordingly, the action potential starts to drop shortly after the opening of the K^+-channels and the repolarization begins [2, 3].

In contrast, the cells of working myocardium, i.e., atrial and ventricular myocardium, have a constant resting potential. As a result, the cells of atria and ventricles cannot be activated by themselves, but can only be passively activated by the neighboring excited cells. The resting potential of the cells in atrial and ventricular myocardium is about $-85\ mV$, which is mainly caused by a high K^+ ion concentration gradient across the membrane ($[K^+]_i/[K^+]_o \approx 50$). When a cell is activated by its neighboring pacemaker cells or working myocardial cells, the ionic balance at rest breaks down: the Na^+ ions flow into the intracellular space resulting in a dramatic increase of the action potential to about $+20\ mV$. Thus, the depolarization of the cell takes place. The duration of depolarization is $1-10\ ms$. Afterwards, the potential remains approximately constant in a plateau phase ($200-250\ ms$). During the plateau phase the cell is not able to respond to additional stimuli. Hence, this phase is also called absolute refractory period. In the subsequent reploarization phase of circa 100 *ms* the action potential of the cell recovers to its resting level due to the outflow of K^+ ions. The cell is in a relative refractory state in this period, i.e., the cell is able to be activated again, but with a shortened duration and a reduced amplitude in comparison to a normal activation. At the beginning of the resting phase the ionic balance is restored very quickly with the aid of the *Na-K* pumps [2, 3]. A typical action potential curve of a working myocardial cell and the whole process are illustrated in Fig. 2.3. In addition to Na^+ and K^+, Ca^{2+} and Cl^- also partially contribute to the genesis of action potential. For the sake of simplicity they are not included in the description above.

Moreover, action potential curves vary in duration and shape for the cells at different depths in the ventricular wall. In Fig. 2.4 the measured action potential curves for epicardial, midmyocardial and endocardial cells in ventricles are shown for different heart rates [4].

2.2.2 Excitation Propagation in the Heart

The excitation in the heart is normally initiated by a pacemaker, whose cells are able to generate excitation spontaneously. In general, pacemaker includes the SA node and the components in the excitation conduction system, i.e., the atrioventricular node (AV node), the bundle of His, Tawara branches and Purkinje fibers (see Fig. 2.2). They have different rhythms if left alone (see Table 2.1). In normal case the SA node has the fastest rhythm among all pacemakers. Therefore, the SA node is called the primary pacemaker and its rhythm determines the activation frequency of the heart. But when the frequency of the other pacemakers is enhanced, the sinus rhythm is depressed or the conduction from the SA node is interrupted, the intrinsic rhythm of the lower components will take place of the sinus rhythm (secondary pacemaker or ectopic pacemaker) [2].

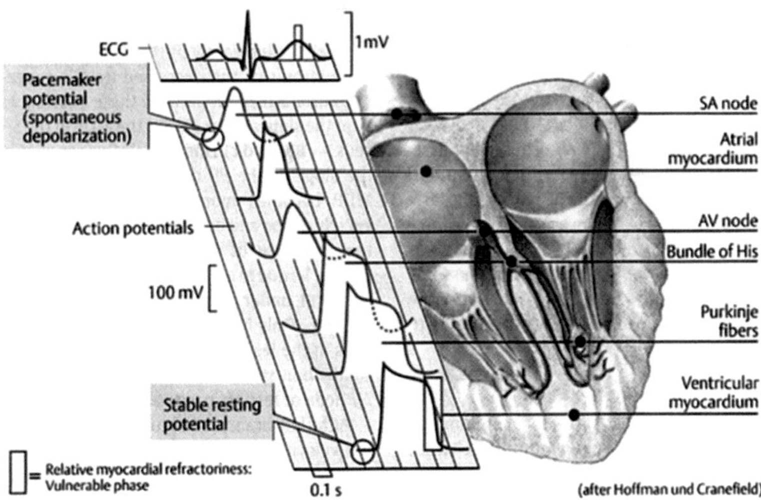

Fig. 2.2. Action potentials of different tissues in the heart [2].

In a normal cardiac cycle the rhythmic impulse starting from the SA node propagates at first from the right atrium to the left atrium. Then, the excitation goes slowly through the AV node, the only passage from atria to ventricles. It ensures that the contraction of atria can finish before the contraction of ventricles. Subsequently, the excitation enters a high speed tree-like network, which is composed of the bundle of His, Tawara branches and Purkinje fibers. At the end of the tree-like network the excitation reaches numerous sites on the endocardium simultaneously. Afterwards, the ventricles become excited from inside to outside and from apex to base. After the plateau phase the ventricles begin to repolarize. The repolarization sequence is determined not only by the depolarization sequence but also by the heterogeneity of action potential duration within the ventricular myocardium. The excitation conduction velocities as well as the intrinsic frequencies of different components of the heart are listed in Table 2.1 [2]. Moreover, the mechanical contraction of the heart is followed by the electrical excitation of myocardium.

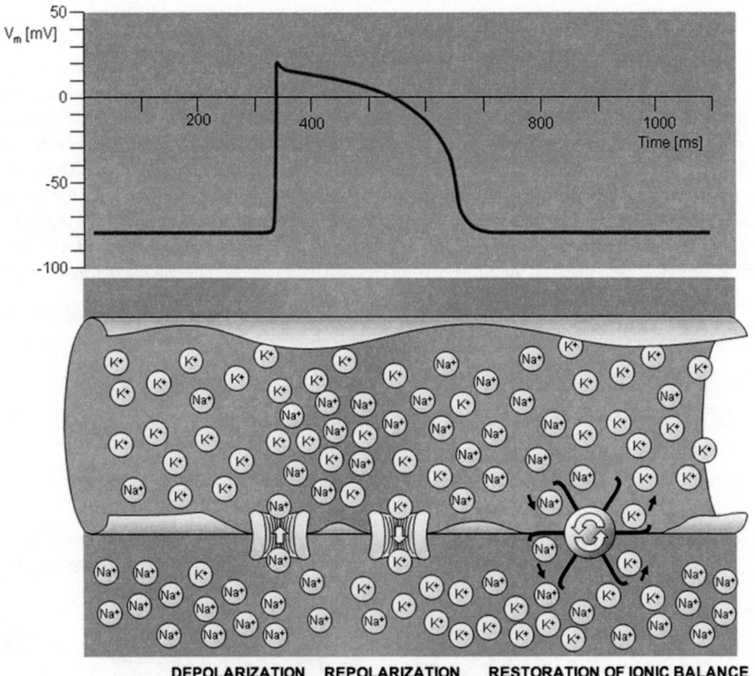

Fig. 2.3. Electrophysiology of active cells [3].

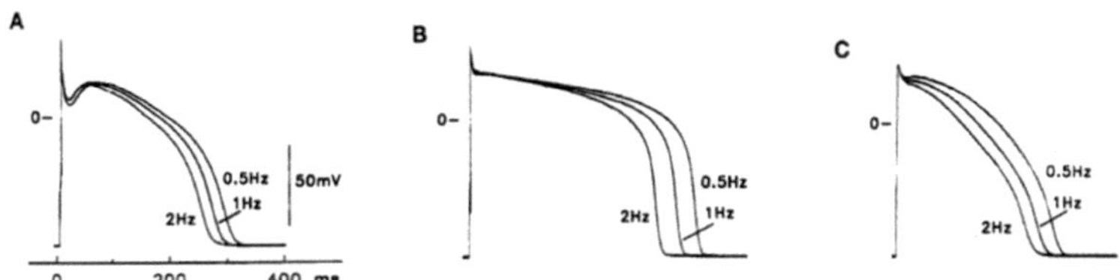

Fig. 2.4. Action potentials recorded from regional cell types in the human right ventricle: epicardial cell (A), midmyocardial cell (B), endocardial cell (C) [4].

Location in the heart	Event	Time [ms]	ECG-terminology	Conduction velocity [mm/s]	Intrinsic frequency [min^{-1}]
SA node	Impulse generation	0		50	70-80
Atrium right	Depolarization	50	P	800-1000	
Atrium left	Depolarization	85			
AV node	Arrival of impulse	50	P-Q interval	20-50	40-55
	Departure of impulse	125			
Bundle of His	activated	130		1000-1500	
Bundle branches	activated	145		1000-1500	20-40
Purkinje fibers	activated	150		3000-3500	
Endocardium					
Left ventricle	Depolarization	175			
Right ventricle	Depolarization	190	QRS	300 (axial) 800 (transverse)	
Epicardium					
Left ventricle	Depolarization	225			
Right ventricle	Depolarization	250			
Epicardium	Repolarization	400			
Ventricles			T	500	
Endocardium	Repolarization	600			

Table 2.1. Electrical events in the heart during cardiac impulse spreading [2, 3].

2.3 Electrocardiography

The electrocardiogram (ECG) registers potential differences on the body surface that are provoked by electrical excitation of the heart. ECG is a remote reflection of the cardiac activities. It provides important information about the heart, e.g., heart position, heart rhythm, impulse propagation. In 1887 the British physiologist Augustus D. Waller made the first ECG measurement on the human body. Later, the Dutch doctor and physiologist Willem Einthoven developed a new string galvanometer in order to overcome the technical difficulties in the measurement of the weak ECG signals at the mV level. Thus, the first practical ECG machine was invented at the beginning of the 20th century. Willem Einthoven received the Nobel Prize in Medicine in 1924 for this invention. Since then, ECG has become a prime tool for the screening and diagnosis of cardiovascular diseases in clinic and practice. However, ECG does not assess the heart status directly. The interpretation of ECG requires an experienced cardiologist and is quite empirical [5].

2.3.1 Measurement System

2.3.1.1 12 Lead ECG

The 12 lead ECG is widely used in today's clinical practice. It consists of three bipolar limb leads, six precordial leads and three unipolar augmented limb leads as illustrated in Fig. 2.5.

The first three standard limb leads (Einthoven leads I, II, III) are introduced by and named after Willem Einthoven. These three bipolar leads are defined as following (see Fig. 2.5 a):

$$V_I = \Phi_L - \Phi_R$$
$$V_{II} = \Phi_F - \Phi_R$$
$$V_{III} = \Phi_F - \Phi_L$$

where Φ_L is the potential at the left arm, Φ_R is the potential at the right arm and Φ_F is the potential at the left foot.

In the early 1930's Frank Norman Wilson suggested an "indifferent electrode" by joining the wires from the right arm, left arm and left foot with 5 $k\Omega$ resistors as the reference for unipolar potential measurement (see Fig. 2.5 b). This reference is later called "Wilson Central Terminal". The mathematical formulation of Wilson Central Terminal is:

$$\Phi_{CT} = \frac{\Phi_L + \Phi_R + \Phi_F}{3}$$

In 1938 the American Heart Association and the Britisch Cardiac Society recommended six unipolar precordial leads applying Wilson Central Terminal as reference (see Fig. 2.5 b). These six leads are named Wilson leads $V_1 - V_6$.

Moreover, Wilson suggested three unipolar limb leads V_R, V_L and V_F. Based on these, Emanuel Goldberger developed three standard unipolar augmented limb leads by increasing the voltage by 50% in comparison to the original version (see Fig. 2.5 c). In Goldberger's suggestion one lead is utilized as the different electrode, while the other two limbs are connected and serve as reference. The three Goldberger leads are defined as follows:

$$aV_R = \Phi_R - \frac{\Phi_F + \Phi_L}{2}$$
$$aV_L = \Phi_L - \frac{\Phi_F + \Phi_R}{2}$$
$$aV_F = \Phi_F - \frac{\Phi_L + \Phi_R}{2}$$

2.3.1.2 Holter Monitor

A Holter monitor is a wearable ECG measurement system developed for long-term ECG recording. It is invented by the American biophysicist Norman Holter in 1949. A Holter monitor normally has three to eight electrodes attached to the patient's chest. The electrodes are connected to a portable recording device, in which the ECG signals are saved. The recording time is usually at least 24 hours. The ECG data are then transferred to a computer for further processing and

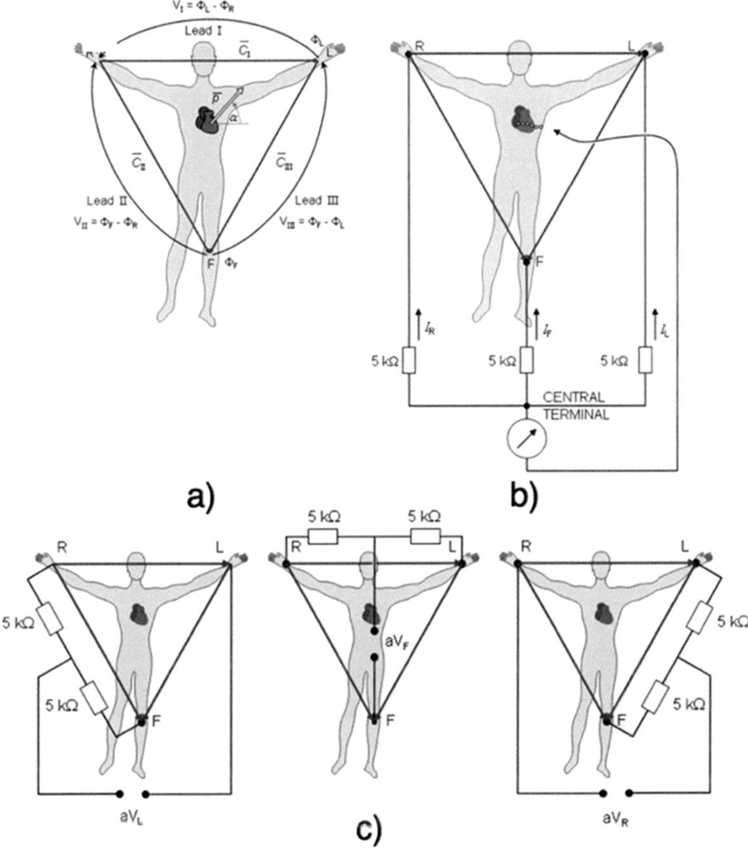

Fig. 2.5. 12 lead ECG measurement system: Einthoven leads (a), Wilson leads (b), Goldberger leads (c) [3].

analysis. The statistics extracted from the long-term ECG signals are helpful for observing and diagnosing cardiac arrhythmias. Because of the wearability of the Holter monitor, it is widely deployed in telemedicine. A Holter monitoring system is shown schematically in Fig. 2.6.

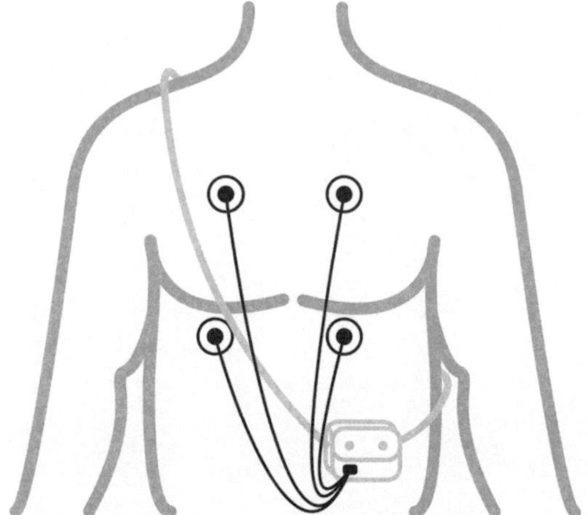

Fig. 2.6. Illustration of a Holter ECG monitoring system.

2.3.1.3 Body Surface Potential Mapping System

Obviously, due to the limited number of recording electrodes the conventional 12 lead ECG is not capable of providing a sufficient spatial resolution for a complete coverage of the electrical signals on the body surface. The body surface potential mapping system possesses a great amount of electrodes (normally $30-256$ electrodes) to cover the patient's thorax surface. Thus, a body surface potential map (BSPM), i.e., a complete distribution of body surface potentials, can be obtained. The high spatial resolution offered by BSPM fulfills the need for such applications like the inverse problem of electrocardiography. In the investigations included in this thesis a 64-channel BSPM system manufactured by BioSemi (Amsterdam, Netherlands) is utilized. The electrode positions are determined using a Polhemus Fastrack electromagnetic localizer (Polhemus Inc, Vermont, USA). A photo captured during the BSPM measurement on a patient is shown in Fig. 2.7.

2.3.2 Notation of Electrocardiogram

A typical ECG describes electrical activities during a cardiac cycle using the following peaks, waves, segments and intervals according to the QRST notation introduced by Willem Einthoven (see Fig. 2.8) [2, 5]:

- P wave ($< 0.3 \ mV$, $< 0.1 \ s$) is associated with the atrial depolarization.
- PQ segment connects P wave and QRS complex.

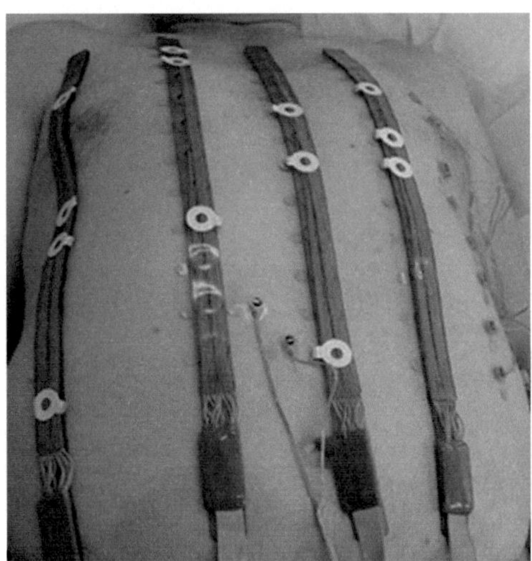

Fig. 2.7. A 64-channel body surface potential mapping system.

- PQ Interval ($< 0.2\ s$) is measured from the beginning of P wave to the beginning of Q peak and represents the time required for atrioventricular conduction.
- QRS complex ($< 0.1\ s$) consists of Q peak, R peak and S peak ($R + S > 0.6\ mV$). QRS complex reflects the depolarization of ventricles. The repolarization of atria is normally not visible on ECG because it tends to be masked by QRS complex.
- ST segment (ca. $0.35\ s$) connects QRS complex and T wave. During ST segment the ventricles are totally depolarized resulting in null signal in ECG. Elevation or depression of ST segment indicate myocardial ischemia or infarction.
- T wave represents the repolarization of ventricles.
- QT interval ($0.35 - 0.4\ s$) indicates the interval between the beginning of Q peak and the end of T wave. It corresponds to the overall time required for the ventricular depolarization and repolarization.
- U wave is a small wave followed by T wave, which is not always seen in ECG. Until now, the genesis of U wave is still under discussion.

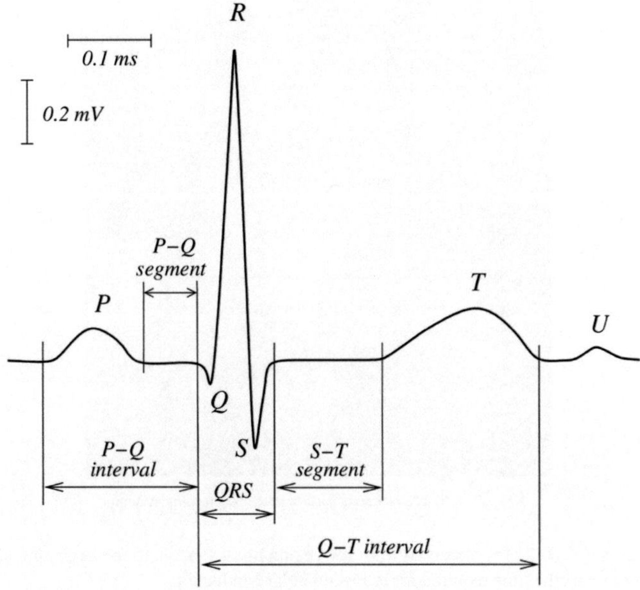

Fig. 2.8. Notation of a normal ECG [6].

Part II

Methods

3

Cardiac Modeling

3.1 Anatomical Model

A computer model that contains the information about the patient's anatomical information and the physical properties of different tissues is the prerequisite for the simulation of cardiac activities and of the forward calculation of ECG, as well as for the further applications based on the computer simulation. In the investigations included in this thesis one standard anatomical model based on the Visible Human dataset and two personalized anatomical models built from the MRI datasets of two patients are deployed. Two types of mesh are utilized: a voxel-based mesh is used for the simulation of cardiac activities and a tetrahedron-based mesh is used in the ECG simulation, the solution of the inverse problem and other relevant applications.

The Visible Human Project is run by the U.S. National Library of Medicine [7]. In this project two datasets were acquired from a male and female cadaver, which were photographed when being cut into slices. The Visible Human datasets are cross-sectional photographs of the human body with high resolution and present the human anatomy in high details. In this thesis the male dataset is used. The dataset is segmented and classified to more than 40 tissue classes. The main segmentation methods applied are region growing and interactive deformable contours [8]. From the segmented data a voxel-based model is built with a resolution of $1\ mm \times 1\ mm \times 1\ mm$ in the heart and a resolution of $2\ mm \times 2\ mm \times 2\ mm$ in the rest of torso. In Fig. 3.1 a cross-sectional image from the Visible Human dataset, the corresponding slice of the data after segmentation and classification, the thorax model and the heart model are shown.

Two individual anatomical models of two patients are also deployed in the investigations. The models are built from magnetic resonance imaging (MRI) scans with the heart in diastolic state, which are made on Siemens Magnetom Vision MRI-scanner at the University of Würzburg, Germany. The torso MRI-scans of Patient 1 have a resolution of $4\ mm \times 4\ mm \times 4.69\ mm$ and Patient 2 $4\ mm \times 4\ mm \times 4\ mm$. Especially for the heart the so-called short axis MRI-scans are made with higher resolutions in comparison to the MRI-scans made for the entire torso. For Patient 1 the short axis MRI-scans of the heart are acquired with a resolution of $2.26\ mm \times 2.26\ mm \times 4\ mm$ and for Patient 2 $2.17\ mm \times 2.17\ mm \times 4\ mm$. First, the 2D MRI-data are converted into a 3D structured dataset. Second, the median filter is performed on the 3D image data to improve the image quality. Then, the dataset is segmented using region growing and interactive deformable contours [8]. Afterwards, the classified MRI data-set is converted into a voxel-based model with an isotropic resolution of $2\ mm \times 2\ mm \times 2\ mm$. The heart model has a resolution of $1\ mm \times 1\ mm \times 1\ mm$.

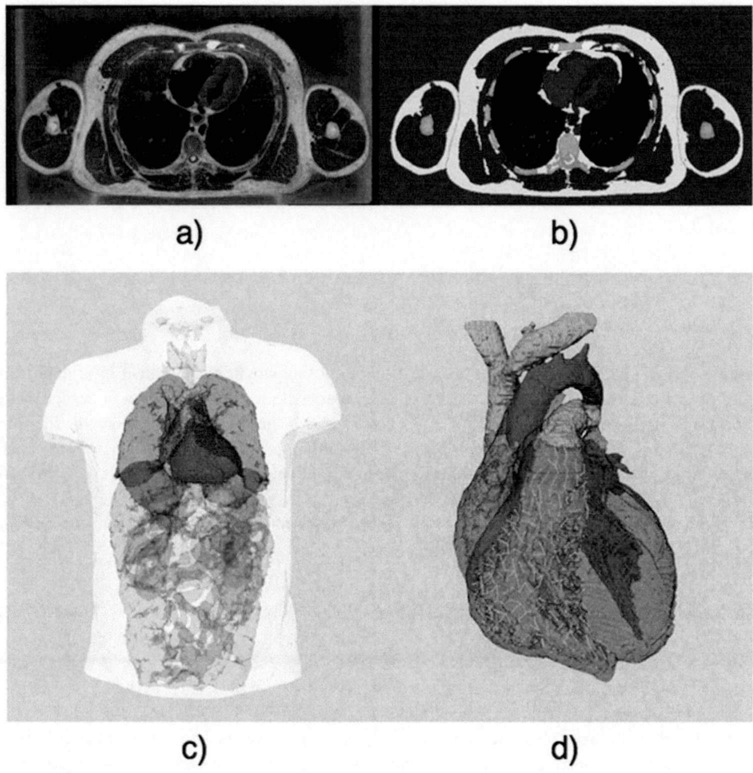

Fig. 3.1. Anatomical model built from the Visible Human dataset: a cross-sectional image from the original Visible Human image data (a), the segmented and classified data (b), the thorax model (c), and the heart model (d) [7, 6].

The torso MRI-data, a short axis MRI-scan, the torso model and the heart model of Patient 1 are shown in Fig. 3.2 and those of Patient 2 are displayed in Fig. 3.3. As can be observed in Fig. 3.2 c and Fig. 3.3 c, atria are difficult to be recognized in the current MRI-data. In addition, the investigations, in which the personalized anatomical models are applied, focus on the cardiac activities in the ventricles. Hence, the atria are only schematically illustrated in the model and not involved in the ECG simulation. Accordingly, no P-wave is present in the simulated ECG using these two models.

In oder to simulate the realistic excitation propagation in the heart, the excitation conduction system is introduced into the heart model, i.e., AV node, Purkinje fibers, Tawara bundles and fascicles. The fiber orientation in the ventricular wall is also built to the model [8]. Moreover, the venous and arterial blood is filled to the right and left ventricles, respectively.

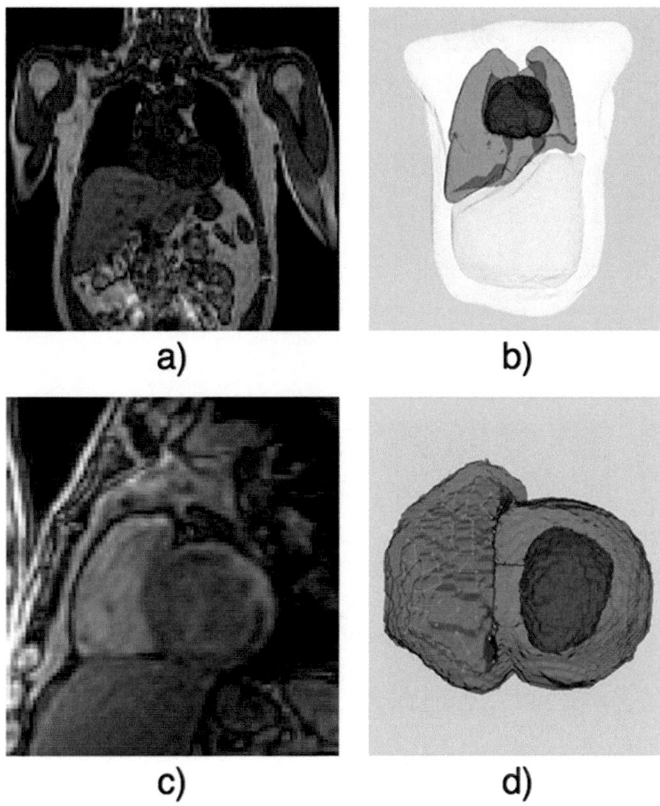

Fig. 3.2. The anatomical model built from the MRI-data of Patient 1: the whole body MRI data (a), the thorax model (b), the short axis MRI data of the heart (c), the heart model (d).

For the forward ECG simulation the voxel-based mesh is converted into a tetrahedron-based mesh, because the unstructured grid like tetrahedron is able to represent complex geometry, e.g., the human body including different tissues, with a relatively small number of elements comparing to the regular structured grid like cubic voxel. The generation of the tetrahedron-based mesh is implemented using the Bowyer-Watson algorithm [9, 10, 11]. Afterwards, adaptive mesh refinement [12, 13] is applied to the tetrahedron-based mesh to improve the accuracy of the further ECG simulation. The conductivity values are assigned to different tissue classes in the volume conductor according to the study published in [14, 15, 16]. Moreover, the unequal anisotropy ratio is modeled in the ventricles: the anisotropy ratio of conductivity tensor (along relative to across the fiber orientation) in extracellular space is set to 3 and that in the intracellular space was set to 9.516. It

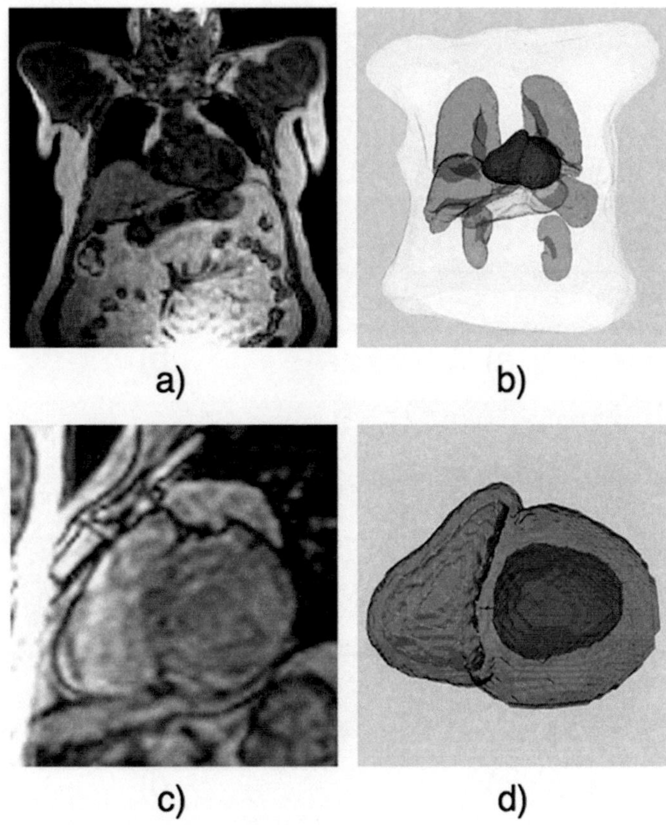

Fig. 3.3. The anatomical model built from the MRI-data of Patient 2: the whole body MRI data (a), the thorax model (b), the short axis MRI data of the heart (c), the heart model (d).

is reported that the unequal anisotropy ratio has remarkable effect on the simulated ECG during T-wave [17].

Furthermore, a dynamic computer model is developed to incorporate the heart motion and respiratory effect into the ECG simulation. It is implemented on the Visible Human model through mesh deformation. The corresponding nodes in the finite element mesh are moved appropriately to mimic the heart motion and the volume change of lungs [18, 19]. From the beginning of a cardiac cycle to the end of QRS-complex the ventricles are in the diastolic state. The ventricles begin to contract at beginning of ST-segment and the minimal volume (end-systolic state) is reached in the middle of T-wave. After that, the ventricles go back to the diastolic state. During the respiration, the conductivity of the lungs varies with the volume change. The conductivity of lungs is set at 0.20279 S/m in the deflated state and at 0.038904 S/m in the inflated state [14, 15, 16]. The conductivity values of the lungs for the time in between are obtained by linear interpolation. The heart model in the diastolic state and in the end-systolic state and the lung model in the deflated state and in the inflated state are shown in Fig. 3.4.

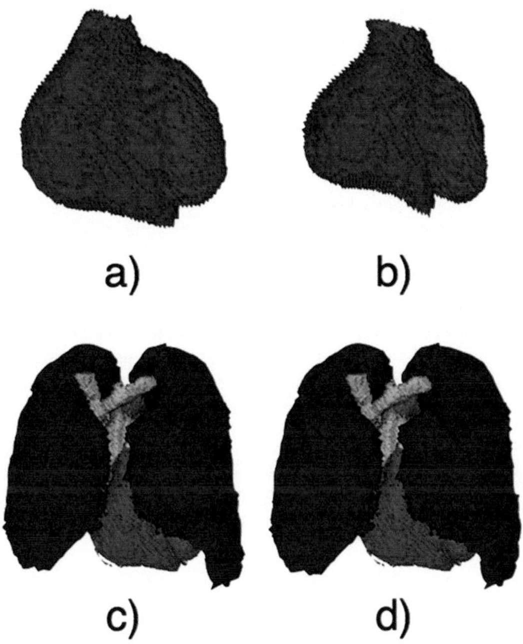

Fig. 3.4. The heart model in the diastolic state (a) and in the end-systolic state (b) and the lung model in the deflated state (c) and in the inflated state (d). The heart is also shown in (c) and (d).

3.2 Cellular Automaton

In this thesis the electrical excitation propagation in the heart is simulated using a rule-based cellular automaton developed at the Institute of Biomedical Engineering, Karlsruhe Institute of Technology (KIT) [8, 20]. Instead of considering the electrophysiology of cardiac cells directly using a cardiac cell model at the microscopic level, which can be extremely time and memory consuming, the cellular automaton runs at a "macroscopic" level on a voxel-based heart model. Every voxel in the model represents a patch of cardiac cells of the same electrophysiological characteristics. By applying the automaton approach a balance point between the computing performance and the simulation accuracy is reached.

In the cellular automaton each voxel is characterized by the following properties:

- the excitability,
- the excitation conduction velocity,
- the fiber orientation (for myocardium only),
- the location in the myocardium (for ventricular myocardium only),
- the intrinsic frequency (for pacemaker only),
- a set of action potential curves at various stimulation frequencies.

In a normal cardiac cycle the electrical impulse is started from the voxels corresponding to the SA node with a given intrinsic frequency. Then, the impulse propagates through the excitable voxels, i.e., myocardium and excitation conduction system. Please note that no excitation conduction between atria and ventricles is allowed except through the AV node. An excitable voxel or a voxel in the excitable state (not in the absolute refractory phase) can be activated by the neighboring excited voxels. After the activation the development of transmembrane voltage in the current voxel compiles with the action potential curve of the corresponding tissue class at the given frequency. The automaton approach does not calculate the action potential curve during the simulation, but uses the action potential curves saved in a library, in which the action potential curves are created beforehand using the ten Tusscher's cardiac cell model [21] while embedding the cells into a virtual wedge preparation. In the ventricular myocardium the transmural heterogeneity is introduced to define the dispersion of action potential curve in the ventricular myocardium (see Fig. 2.4). A schematic representation of the working principle of the cellular automaton is shown in Fig. 3.5.

The simulated cardiac cycle is 1 s. The simulation is performed in a voxel-based heart model with a spatial resolution of 1 $mm \times 1\ mm \times 1\ mm$ and with a time step of 1 ms. As output, the simulated transmembrane voltages in the heart model are saved every 4 ms.

3.3 Modeling of Cardiac Diseases

In the present thesis two cardiac diseases, i.e., premature ventricular contraction and myocardial infarction, are considered besides the normal heart beat. They are modeled and simulated using the cellular automaton.

3.3.1 Premature Ventricular Contraction

Premature ventricular contraction (PVC) is a common cardiac arrhythmia. The causes of PVC can be cardiac ischemia, heart attack, lack of oxygen, cardiomyopathy or medications. During a PVC the electrical excitation and contraction of ventricles is initialized by an ectopic center in the ventricular myocardium. In a normal heart beat the electrical excitation starts from the AV node in atria and reaches the entire ventricular endocardium through the excitation conduction system. In this way, both ventricles contract almost simultaneously in the normal case. In the case of PVC, the excitation starting from the ectopic center propagates slowly through the myocardium from one ventricle to the other. Therefore, the contraction of ventricles is not synchronized and results

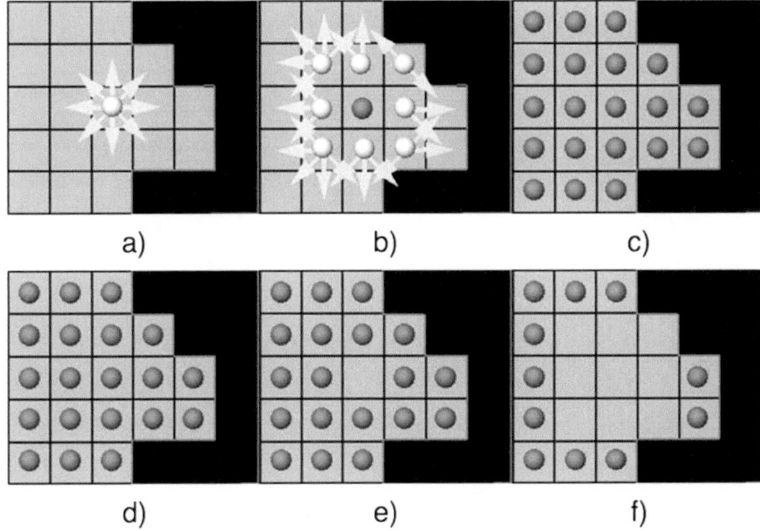

Fig. 3.5. Neighbor propagation of a cellular automaton on a surface. Empty squares represent resting cells. Each excited cell (light cycle), communicates the excitation to its eight neighbors (arrow) (a, b). After an explicit time interval the cell switches to an unexcitable state (dark cycle) (c, d). Finally, the cell returns back to an excitable cell (e, f). This process proceeds along the given geometry [22].

in an inefficient circulation. As illustrated in Fig. 3.6 b, PVC normally leads to a QRS-complex of high amplitude and a prolonged cardiac cycle. Infrequent PVCs usually do not need any treatment. However, a high occurrence of PVC may possibly cause ventricular tachycardia and may lead to cardiomyopathy on the long run. Radiofrequency catheter ablation can be applied to eliminate the trigger of PVC. In the cellular automaton PVC is modeled by applying an external stimulus in the ventricular myocardium. It is defined by 4 parameters: three coordinates of the ectopic center and the time instant when the PVC starts.

3.3.2 Myocardial Infarction

Myocardial infarction is a leading cause of morbidity and mortality. It is caused by the shortage of oxygen supply to part of the myocardium, e.g, when a coronary vessel is blocked. In early stage myocardial ischemia emerges. In the ischemic region the cardiac tissue bears an insufficient oxygen supply and the cardiac cells show a restricted vitality, i.e., a reduced excitation propagation velocity and an action potential with decreased amplitude and length. If this condition persists, myocardial infarction will develop. It leads to necrosis of the myocardium. In the necrotic region the gap junctions between cells are broken. The necrotic tissue is non-excitable and no electrical current sources exist. A transition zone surrounds the necrotic center, in which a gradual change from the necrotic tissue to the healthy tissue is shown. In Fig. 3.7 a myocardial infarction caused

NORMAL SINUS RHYTHM
Impulses originate at S-A node at normal rate

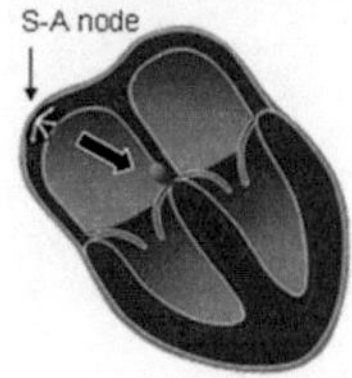

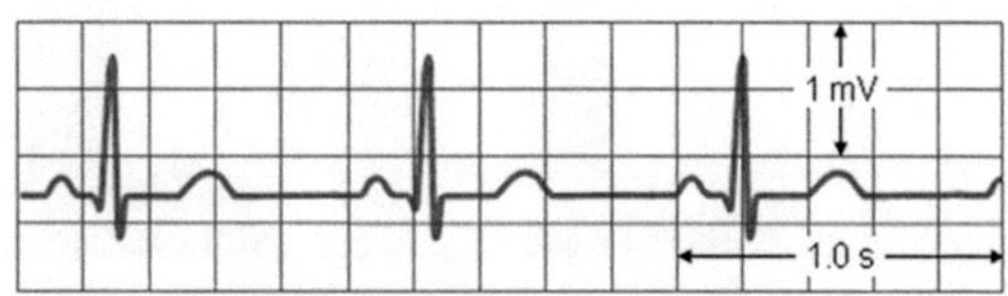

All complexes normal, evenly spaced. Rate 60 – 100/min.

a)

PREMATURE VENTRICULAR CONTRACTION
A single impulse originates at right ventricle

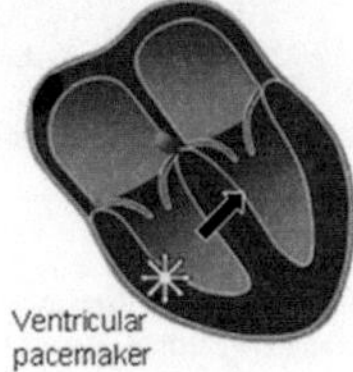

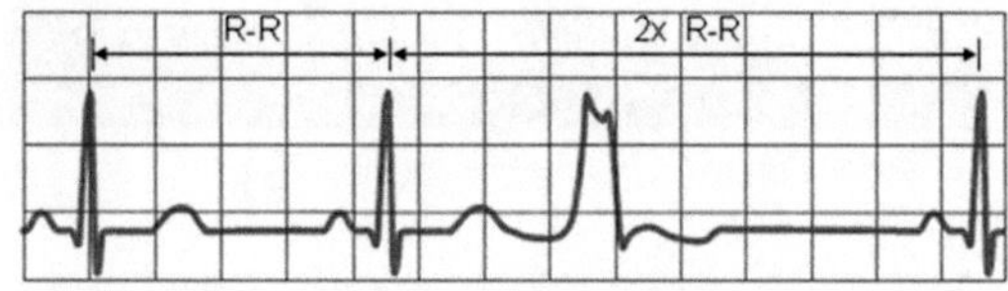

Time interval between normal R peaks is a multiple of R-R interval

b)

Fig. 3.6. Comparison between the ECGs of normal sinus rhythm (a) and of PVC (b) [3].

by the occlusion of the left coronary artery is illustrated. Myocardial infarction may lead to immediate sudden cardiac death. In case the patient survives the scar tissue in the region of healed myocardial infarction is still a potential trigger for life threatening arrhythmias.

For diagnosis and therapy of myocardial infarction it is of importance to detect and localize myocardial infarction efficiently and effectively. In the clinical practice the routine tool to diagnose myocardial infarction is the standard 12 lead ECG. For this purpose, guidelines were proposed for the standardization and interpretation of the 12 lead ECG [24, 25]. Some tomographic techniques are capable of imaging myocardial infarction, e.g., late enhancement MRI and myocardial perfusion scintigraphy. However, they can only show the metabolistic but not the electrophysiological consequences of the myocardial infarction. In addition, blood test provides another approach to determining myocardial infarction by checking biomarkers of ischemia and infarction [26, 27], but with a delay of several hours.

In order to understand the myocardial infarction from an electrophysiological point of view, several investigations have been done to model the behavior of ischemic and infarcted cardiac tissue in the recent years [28, 29, 30]. At the same time, much research has been taken out to study the

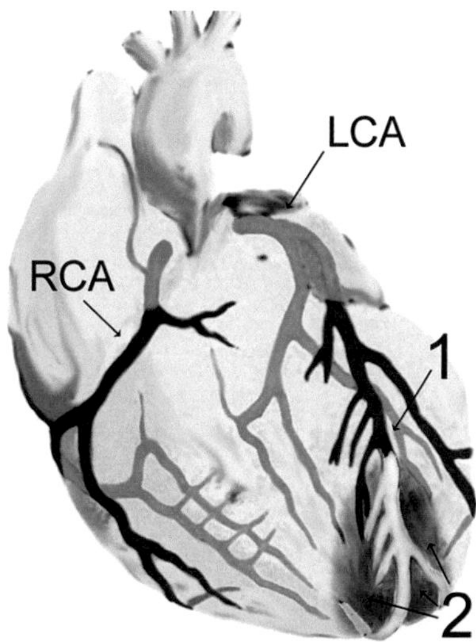

Fig. 3.7. Diagram of a myocardial infarction (2) of the tip of the anterior wall of the heart (an apical infarct) after occlusion (1) of a branch of the left coronary artery (LCA, right coronary artery = RCA) [23].

effects of myocardial ischemia or infarction in various regions of the heart on ECG and on the body surface potential signals by computer simulation [31, 32, 33, 34].

In the cellular automaton myocardial infarction is modeled by changing the local parameters for the excitation amplitude and propagation velocity of the affected voxels using the function shown in Fig. 3.8. In the necrotic center the excitation amplitude and the propagation velocity of cells are set to zero. The parabolic curve represents the incremental vitality of cardiac cells in the transition zone, from the necrotic center to the healthy tissue. In the modeling a myocardial infarction is defined by 5 parameters: three coordinates of the infarction center, the size of the affected area in *mm* and the slope of the parabola used to describe the transition zone [20].

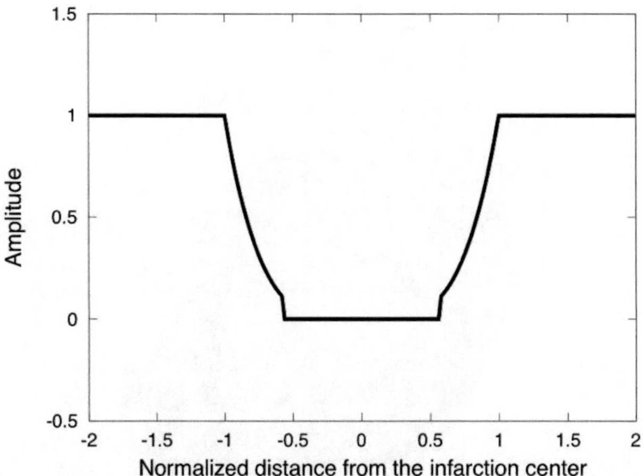

Fig. 3.8. The dependency of excitation amplitude and propagation velocity from the distance to the center of myocardial infarction. The unit length 1 in x-axis indicates the radius of infarction area.

4

Forward Problem of Electrocardiography

4.1 Introduction

The objective of the forward problem of electrocardiography is the calculation of the electrical potentials on the body surface or on the heart surface from a given distribution of bioelectrical sources in the heart. As input to the forward problem, the distribution of cardiac sources can be provided either by invasive measurement or by computer heart model. The system is defined by the volume conductor model that contains the anatomical information of the patient and the physical properties of different tissues in the human body. As outcome, the simulated electrical potentials on the body surface are ECG signals and those on the heart surface are epicardial potentials, the so-called electrogram.

The common applications of the forward problem of ECG include:

- investigation of the effect of electrophysiological properties of different tissues on the body surface potentials or epicardial potentials,
- validation of the cardiac cell models,
- optimization of the ECG measurement systems,
- calculation of the transfer matrix for the inverse problem,
- solution of the inverse problem with the model-based optimization.

Moreover, the methodology of the forward problem can also be deployed to support the development of the cardiac-related diagnostic and therapeutic technologies, e.g., determination of the optimal positioning and size of the defibrillation electrodes as well as of the optimal shock duration for the defibrillator. One topic included in the present thesis is the development of an impedance-based catheter positioning system for cardiac mapping and navigation, in which the principle of the forward problem is also applied (see Chapter 6).

In the present thesis the forward problem is considered linear and quasi-stationary. It starts from the distribution of transmembrane voltages in the myocardium at one certain time instant, which is given by the cellular automaton. The bidomain model is applied to simplify the problem formulation. Finally, the simulated body surface potentials are obtained by solving the problem on the tetrahedron-based volume conductor using the finite element method.

4.2 Bidomain Model

The human heart is composed of billions of excitable cells. Therefore, the time and memory consumption needed for a computation that involves such a great amount of cardiac cells can be extremely high. In order to reduce the computing complexity, the bidomain model was proposed by considering the discrete cells in an averaged continuum at the macroscopic level [35, 36]. From the microscopic point of view, a cardiac cell is surrounded by a plasma membrane that separates the intracellular space of this cell from the extracellular space. Gap junctions provide the intercellular communication between cells, which allow currents to flow from one cell into its neighbors. From the macroscopic point of view, the interconnected intracellular space and the extracellular space can be considered as two continuous domains that fill up the whole space inside the cardiac tissue. The two domains are divided by the membrane at any point in the cardiac tissue and represented with their averaged properties. The principle of bidomain model is illustrated in Fig. 4.1.

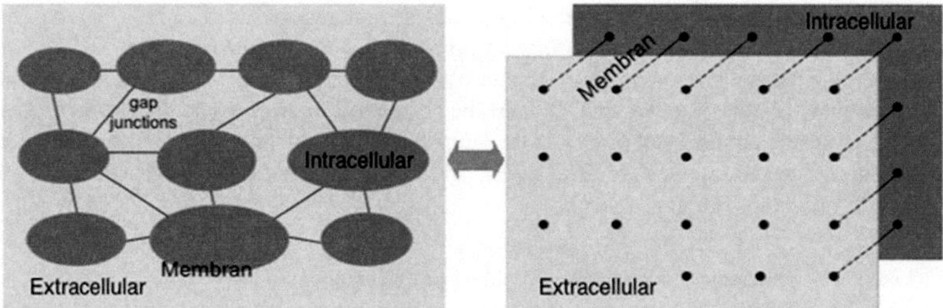

Fig. 4.1. Bidomain model [37]

The current densities in both spaces are given by Ohm's law as

$$\mathbf{J}_i = \sigma_i \mathbf{E}_i + \mathbf{J}_i^{imp} \tag{4.1}$$

$$\mathbf{J}_e = \sigma_e \mathbf{E}_e + \mathbf{J}_e^{imp}, \tag{4.2}$$

where σ_i and σ_e are the intracellular and extracellular conductivity tensors of bidomain model, respectively; $\mathbf{J}_i^{imp}$ and $\mathbf{J}_e^{imp}$ are the impressed current densities in the intracellular and extracellular spaces, respectively.

Since the total current in the cardiac tissue is solenoidal, the intracellular current density $\mathbf{J}_i$ and the extracellular current density $\mathbf{J}_e$ satisfy the equation below

$$\nabla \cdot (\mathbf{J}_i + \mathbf{J}_e) = 0. \tag{4.3}$$

Because the electrical fields invoked by the bioelectric sources in the active tissue can be generally treated as quasi-stationary [38, 39, 3], the intracellular electrical fields $\mathbf{E}_i$ and extracellular electrical fields $\mathbf{E}_e$ are given by

$$\mathbf{E}_i = -\nabla \Phi_i \tag{4.4}$$

$$\mathbf{E}_e = -\nabla \Phi_e, \tag{4.5}$$

where Φ_i and Φ_e indicate scalar potentials in the intracellular and extracellular spaces, respectively.

Substituting equations 4.1, 4.2, 4.4 and 4.5 into Eq. 4.3 gives

$$\nabla \cdot (\sigma_i \nabla \Phi_i + \sigma_e \nabla \Phi_e) = 0. \tag{4.6}$$

By introducing the transmembrane voltage

$$V_m = \Phi_i - \Phi_e, \tag{4.7}$$

the intracellular potential can be excluded in Eq. 4.6:

$$\nabla \cdot ((\sigma_i + \sigma_e)\nabla \Phi_e) = -\nabla \cdot (\sigma_i \nabla V_m). \tag{4.8}$$

Because of the continuity of the extracellular potential at the boundary between the cardiac tissue and other tissues as well as the disappearance of the intracellular current outside the cardiac tissue, the bidomain model can be extended into the whole human body with $\sigma_i = 0$ for the tissues other than cardiac tissue. Thus, Eq. 4.8 is valid in the entire torso model including the cardiac tissue and other tissues.

The right hand term of Eq. 4.8 can be interpreted as the impressed current source density

$$I_{sv} = \nabla \cdot (\sigma_i \nabla V_m), \tag{4.9}$$

which arises from the bioelectric activity of active tissue due to the conversion of energy from the chemical form to the electrical form. Using Eq. 4.9 the impressed current source density I_{sv} can be calculated from transmembrane voltages V_m.

Eq. 4.8 is further formulated as a Poisson's equation

$$\nabla \cdot (\sigma \nabla \Phi) = -I_{sv} \tag{4.10}$$

with the bulk conductivity $\sigma = \sigma_i + \sigma_e$ and Φ is the extracellular potential defined in the entire space [6].

4.3 Solution of the Forward Problem

The forward problem of electrocardiography incorporating bidomain model is formulated as follows:

$$\nabla \cdot (\sigma \nabla \Phi) = -I_{sv} \quad \text{in } \Omega \tag{4.11}$$

with the boundary conditions

$$\Phi = \Phi_D \quad \text{on } \Gamma_1 \tag{4.12}$$

$$(\sigma \nabla \Phi) \cdot \mathbf{n} = 0 \quad \text{on } \Gamma_2 \tag{4.13}$$

where Ω is a finite domain and Γ its boundary. Γ_1 and Γ_2 are parts of the boundary, $\Gamma_1 \cup \Gamma_2 = \Gamma$, $\Gamma_1 \cap \Gamma_2 = \varnothing$. Eq. 4.12 describes the Dirichlet boundary condition which refers to the reference electrode, at which the zero potential is defined (Γ_1). The homogenous Neumann condition given by Eq. 4.13 indicates the boundary between the torso surface and air (Γ_2) [6].

At this point, the main task is to solve the Poisson's equation (Eq. 4.11) with sufficient accuracy in an efficient way. In this work the finite element method is selected to solve the problem because of its ability to handle complex geometries and inhomogeneous anisotropic media. This method divides the solution domain into a set of contiguous discrete sub-domains (finite elements) of simple geometrical shapes, e.g., linear tetrahedral elements [40].

Then, the sought potential Φ is discretized in the finite elements and is approximated by

$$\tilde{\Phi} = \sum_{k=1}^{m} \alpha_k(x,y,z) \cdot \Phi_k, \tag{4.14}$$

where Φ_k are the potentials at node k in the volume conductor and α_k denotes the appropriate interpolation functions at node k, which has the following form:

$$\alpha_k = \begin{cases} 1 & \text{at the node } k \\ 0 & \text{at the other element nodes} \end{cases} \tag{4.15}$$

By employing the method of weighted residual [41] a residual results from substituting $\tilde{\Phi}$ into Eq. 4.11:

$$\nabla \cdot (\sigma \nabla \tilde{\Phi}) + I_{sv} = R. \tag{4.16}$$

The optimal approximate $\tilde{\Phi}$ is reached when the residual R is minimized. The integral of the residual over Ω is set to 0 with a set of weighting functions ω_k

$$\int_\Omega \nabla \cdot (\sigma \nabla \tilde{\Phi}) \omega_k \mathrm{d}v + \int_\Omega I_{sv} \omega_k \mathrm{d}v = \int_\Omega R \omega_k \mathrm{d}v = 0 \ , k = 1, 2, \ldots m. \tag{4.17}$$

In the Galerkin formulation the weighting functions ω_k are chosen to be equal to the basis functions α_k

$$\int_\Omega \nabla \cdot (\sigma \nabla \tilde{\Phi}) \alpha_k \mathrm{d}v + \int_\Omega I_{sv} \alpha_k \mathrm{d}v = 0 \ , k = 1, 2, \ldots m. \tag{4.18}$$

By applying the product rule of divergence and the divergence theorem Eq. 4.18 is rewritten as

$$\oint_\Gamma \alpha_k(\sigma \nabla \tilde{\Phi}) \cdot \mathbf{n} \mathrm{d}s - \int_\Omega (\sigma \nabla \tilde{\Phi}) \nabla \alpha_k \mathrm{d}v + \int_\Omega I_{sv} \alpha_k \mathrm{d}v = 0 \ , k = 1, 2, \ldots m. \tag{4.19}$$

In Eq. 4.19 the first term must vanish, because the normal of the current density in the intracellular domain is equal to zero both on Γ_1 (the reference) and on Γ_2 (the torso surface). Thus, Eq. 4.19 is expressed in

$$\sum_{j=1}^{m} \left(\int_{\Omega} (\sigma \nabla \alpha_j) \nabla \alpha_k dv \right) \Phi_j = \int_{\Omega} I_{sv} \alpha_k dv \ , k = 1, 2, \ldots m. \tag{4.20}$$

Further, Eq. 4.20 can be written in the matrix form as

$$S\Phi = b, \tag{4.21}$$

where S is a sparse $m \times m$ coefficient matrix called system matrix, Φ is an $m \times 1$ matrix of the desired unknown electrical potentials at the nodes and b is the so-called right-hand vector that includes the sources. The system matrix is symmetric and satisfies the positive definiteness.

By incorporating the Dirichlet boundary condition (Eq. 4.12) into the finite element approximation of the forward problem, Eq. 4.21 can be formulated in the form of submatrices:

$$\begin{pmatrix} S_{VV} & S_{VD} \\ S_{DV} & S_{DD} \end{pmatrix} \begin{pmatrix} \Phi_V \\ \Phi_D \end{pmatrix} = \begin{pmatrix} b_V \\ b_D \end{pmatrix}, \tag{4.22}$$

where subscripts D and V denote the nodes on the boundary Γ_1 and the other nodes in the volume conductor, respectively. Thus, the subvector Φ_V contains the unknown node variables in the mesh and Φ_D contains the node variables at the reference point.

The linear system (Eq. 4.22) can be solved with various numerical calculation techniques, e.g., direct methods like Gaussian elimination and Cholesky decomposition, and iterative methods like method of steepest descent and conjugate gradient method [42, 43, 44, 45]. Finally, the unknown extracellular potential Φ_V are obtained. Φ_V on the heart surface are the epicardial potentials, Φ_V on the body surface are the body surface potentials. The ECG signals are also derived from Φ_V at the measurement electrodes [6].

4.4 Realistic Environment

In order to set up a realistic test environment that is close to the scenario of the clinical practice for the inverse problem of electrocardiography (see Chapter 5), different sorts of error are added to the ECG simulation. The errors considered in this realistic environment include measurement noise, baseline wander, inaccuracy in the localization of body surface electrodes, the inaccuracy in the estimation of tissue conductivities and the modeling errors introduced by neglecting the heart motion and respiration. This study is performed on the male Visible Human model.

First, the heart dynamics and the respiration are introduced into the anatomical model as described in Section 3.1. In the simulation a respiratory cycle has a period of 4 s, which contains 4 cardiac cycles (1 s for each). The simulation starts from the deflated state, followed by 1.5 s for inspiration, 1.5 s for expiration and at the end 1 s in rest. Within a respiratory cycle of 4 s 1000 states are created with a time step of 4 ms.

Then, the 64 measurement electrodes on the body surface are shifted towards the bottom-right direction by 5 mm (see Fig. 4.2). It represents the error caused by the inaccuracy of electrode localizer arising from an erroneous magnetic localizer or a bad registration. The inaccuracy in the estimation of tissue conductivities is also introduced: the conductivities of myocardium, of blood inside the heart chambers and of lungs in the volume conductor are set to 80% of their original values.

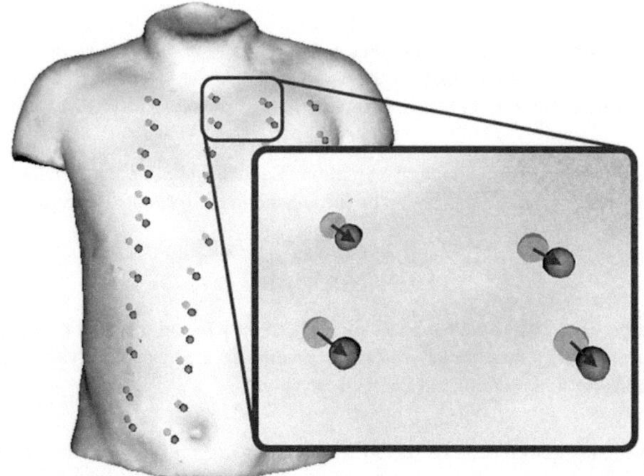

Fig. 4.2. Shifting of the ECG measurement electrodes by 5 *mm* towards the bottom-right direction [46].

Afterwards, the forward ECG calculation is performed on the dynamic model and the 64-channel ECG is recorded at the shifted electrodes. The simulated ECG is then contaminated with an additive white Gaussian noise of 30 *dB* SNR. At the end, a random baseline wander is added to each channel, which has the form of a sinusoidal with a frequency of 0.1 *Hz* and a random amplitude between 0.1 *mV* and 0.2 *mV* as well as a random phase shift in different channels. In this way a multichannel ECG is generated in a simulation environment that resembles the realistic situation.

4.5 Optimization of Electrode Positions for a Wearable ECG Monitoring System

Nowadays, the Holter ECG system with three to eight measurement electrodes (see Section 2.3.1) is widely deployed in recording long-time ECG. But it does not examine the heart comprehensively. The body surface potential mapping system (see Section 2.3.1) provides a complete coverage of the patient's thorax. However, it limits the mobility of the patient due to the great number of measurement electrodes. In this study a wearable ECG system with a minimal number of electrodes (see Fig. 4.3) is proposed, which enables the real-time ECG monitoring on the people at high risk of developing myocardial infarction without disturbing their everyday life. It is also able to detect ischemic events and myocardial infarction from the recorded ECG signals at the very first moment.

The present investigation is aimed to determine an optimal electrode configuration for the proposed wearable ECG monitoring system using the results of the forward simulation of ECG. The

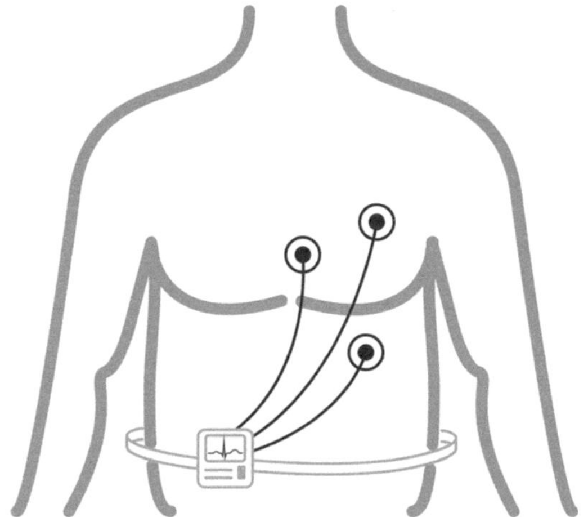

Fig. 4.3. Illustration of the proposed wearable ECG monitoring system including a minimal number of measurement electrodes on the body surface [47].

minimal number of electrodes will be defined and the electrode positions will be optimized in terms of the detection of myocardial infarction. Optimum in this work means: no infarction will be missed.

In the simulation study the two personalized anatomical models built from the patients' MRI data (see Section 3.1) are employed. In order to construct a data base (stochastical basis) including different myocardial infarctions for the optimization, 153 myocardial infarctions of different sizes and at various sites in the entire left ventricular wall and septum are simulated using the cellular automaton. For detailed description of the stochastical basis please see Section 7.3. The corresponding body surface potentials are computed by solving the forward problem of ECG. Because the most significant symptom of myocardial infarction shown in ECG is the elevation or depression in ST-segment, the study focuses on the ST-segment. To consider the entire ST-segment all at once the ST-integral of body surface potentials is calculated

$$\mathbf{b} = \sum_{\kappa=K_1}^{K_2} b_\kappa,$$ (4.23)

where b_κ denotes the body surface potentials at the time instant κ, and $\mathbf{b}$ stands for the integral of the body surface potentials over the time interval between the beginning of ST-segment K_1 and the end of ST-segment K_2, which is also called the ST-integral map in this thesis.

The optimization of electrode positions is based on the analysis of the ST-integrals. At first N electrode positions that are evenly distributed on the surface of torso model are selected ($N = 663$ for Patient 1 and $N = 633$ for Patient 2). $D_k(i, j)$ denotes the potential difference (dipolar ECG signal) between the ith and jth electrodes on the body surface in the case of the myocardial infarction no. k. Then, The ST-integral of $D_k(i,j)$ is calculated as $\mathbf{D}_k(i,j)$. In addition, the ST-integral between the ith and jth electrodes is also calculated for the healthy case (no infarction), denoted as $\mathbf{D}_h(i,j)$. Afterwards, a search procedure is started in order to find an electrode pair that is able to detect all infarctions. The process begins with a random choice of electrodes i and j and infarction $k=1$. If the difference between $\mathbf{D}_k(i,j)$ and $\mathbf{D}_h(i,j)$ is above a threshold T, this pair of electrodes is able to detect the myocardial infarction no. k. Then, the next myocardial infarction no. $k+1$ will be examined by using the ST-integral at the ith and jth electrodes with the current threshold T. If the difference is equal or below the given threshold, this pair of electrodes fails to detect the current myocardial infarction. Then, another combination of i and j will be tested starting again with infarction $k=1$ and the current threshold T. This analysis procedure will run throughout all the 153 myocardial infarctions in the data base and all combinations of 2 electrodes with a given threshold. In case that all the infarctions in the data base can be detected by one pair of electrodes with the given threshold, this pair of electrodes will be the selected electrode configuration for the proposed wearable ECG monitoring system. If it is not the case, the threshold T will decrease, and the optimization process will restart. The entire process will be repeated until the threshold decreases to a very small value that cannot be distinguished from noise [47]. The flow diagram of this process is shown in Fig. 4.4. In case that a single pair of electrodes is not sufficient to detect all the myocardial infarctions, two pairs of electrodes but with one electrode in common, i.e., 3 electrodes, will be used. In this case one myocardial infarction can be considered as being successfully detected when at least one pair of electrodes can detect the current myocardial infarction .

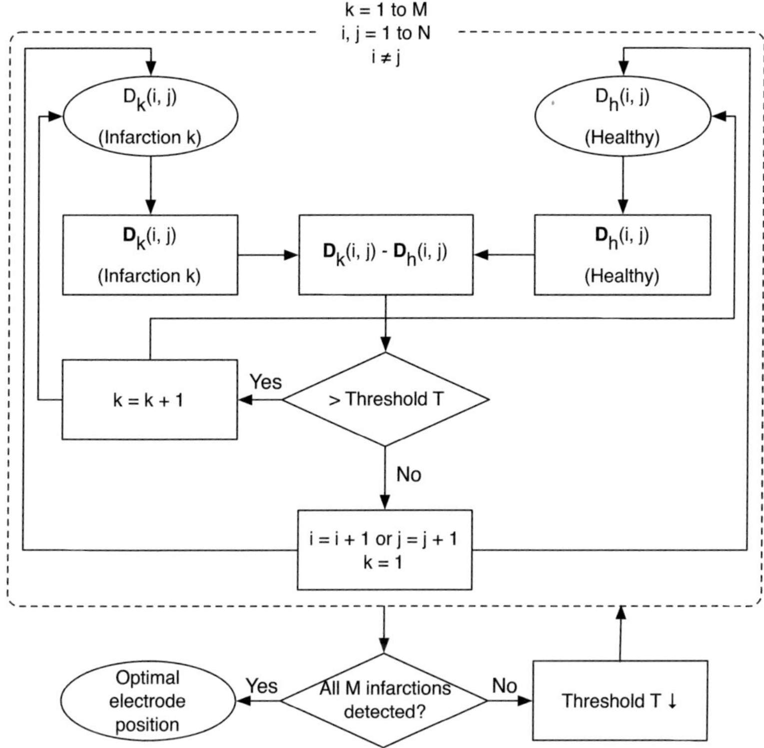

Fig. 4.4. Flow diagram demonstrating the process of determination of the optimal electrode position for the wearable ECG monitoring system to detect cardiac infarction, $M = 153$ for both patients, and $N = 663$ for Patient 1 and $N = 633$ for Patient 2 [47].

5

Inverse Problem of Electrocardiography

5.1 Introduction

The inverse problem of electrocardiography is aimed to reconstruct bioelectrical sources in the heart from the multichannel ECG measured on the body surface. Besides ECG signal the information about patient's anatomy and the knowledge of the physical properties of the human body are also considered in the solution of the inverse problem of ECG. It is a promising noninvasive imaging technique, which provides cardiologists with the useful and straightforward information about the functional status of the heart. However, the following inherent difficulties of the inverse problem of ECG must be handled carefully:

- the non-uniqueness of the inverse solution: different cardiac source distributions may result in the same body surface potentials;
- the underdetermination of the inverse problem of ECG: the number of ECG measurement electrodes is extremely small in comparison to the large number of unknown sources in the heart;
- the ill-posedness of the inverse problem: the inverse solution is very sensitive to small perturbations in the measurement and to the errors in the model.

In order to obtain a stable and physiologically reasonable inverse solution, an appropriate source model should be selected to ensure the uniqueness of the solution, noise depression and baseline removal should be applied to the ECG signal, and the errors in the segmentation and in the modeling should be kept as small as possible. Furthermore, regularization techniques must be deployed to impose additional constraints on the solution.

5.2 Formulation of the Inverse Problem

In the integral form the linear inverse problem of ECG can be described as

$$b(\eta,t) = \int_{\Omega} A(\eta,\xi) x(\xi,t) \, \mathrm{d}\Omega_{\xi}, \tag{5.1}$$

where $x(\xi,t)$ is the source function, $b(\eta,t)$ is the measurement function, and $A(\eta,\xi)$ is the transfer function describing the contribution of a unitary source $\delta(\xi - \xi_0)$ (the Dirac delta function positioned at ξ_0) to the measurement $b(\eta,t)$. Ω_{ξ} denotes the source domain. In the inverse problem the source function $x(\xi,t)$ is to be sought from the given measurements $b(\eta,t)$.

Eq. 5.1 can also be rewritten in the matrix form as

$$Ax = b, \tag{5.2}$$

where x denotes the source vector, y stands for the measurement vector, and A is the transfer matrix describing the relation between the cardiac sources and the measurement signal on the body surface.

The computation of the lead-field matrix is performed by the solution of the forward problem [48]. Please note the following derivation is based on selecting epicardial potentials as source model. The matrix formulation of the forward problem (Eq. 4.21) will be again under consideration. This time Dirichlet boundary condition related to the cardiac sources and Neumann boundary condition defined at the surface of the volume conductor are both applied. Eq. 4.21 is rewritten as follows

$$\begin{pmatrix} S_{TT} & S_{TV} & 0 \\ S_{VT} & S_{VV} & S_{VE} \\ 0 & S_{EV} & S_{EE} \end{pmatrix} \begin{pmatrix} \Phi_T \\ \Phi_V \\ \Phi_E \end{pmatrix} = \begin{pmatrix} 0 \\ 0 \\ 0 \end{pmatrix}. \tag{5.3}$$

The first matrix in Eq. 5.3 is the system matrix, which is subdivided into block matrices. The system matrix describes the relations between three potential vectors Φ_T, Φ_E and Φ_V. The subscripts T, V and E denote torso, volume and epicardium, respectively. Because there are no direct connections between the epicardial nodes and torso nodes, the block matrices S_{TE} and S_{ET} in the system matrix equal zero. Φ_T is the vector of electrical potentials at the nodes on the torso surface, Φ_E is the vector of electrical potentials at the nodes located on the epicardium and Φ_V are the electrical potentials at the rest of the nodes in the volume conductor.

Further, the relation between body surface potentials Φ_T and the epicardial potentials Φ_E is reached:

$$\Phi_T = (S_{TT} - S_{TV} S_{VV}^{-1} S_{VT})^{-1} S_{TV} S_{VV}^{-1} S_{VE} \Phi_E. \tag{5.4}$$

Introducing

$$A = (S_{TT} - S_{TV} S_{VV}^{-1} S_{VT})^{-1} S_{TV} S_{VV}^{-1} S_{VE} \tag{5.5}$$

into Eq. 5.4 gives

$$\Phi_T = A \Phi_E \tag{5.6}$$

Since Eq. 5.6 is identical to Eq. 5.2, A defined in Eq. 5.5 is the transfer matrix to be sought. Because S_{VV} includes almost all the nodes of the torso model except the nodes on the heart surface and on the torso surface, the calculation of S_{VV}^{-1} is very time and memory consuming. For this reason, another approach is used to compute the transfer matrix A.

Considering a unit source vector x_i, which is defined as

$$x_i = \begin{cases} 1 & \text{at the node } i \\ 0 & \text{at the other nodes.} \end{cases} \tag{5.7}$$

The ith column of the transfer matrix A can be interpreted as the measurement signal invoked by a single unit source at the ith node in the source domain:

$$a_i = Ax_i. \tag{5.8}$$

Applying Eq. 5.8 to every node in the source domain the entire transfer matrix A is obtained. Moreover, this approach satisfies the integral formulation of the inverse problem (Eq. 5.1). Therefore, it can be applied for any source model under the assumption that the inverse problem of ECG is linear [6].

5.3 Source Model

At an early stage of the research basic source models, e.g., single current dipole or multipole and multiple current dipoles, were deployed to reduce the complexity of the inverse problem of ECG [3]. Nowadays, a popular source model is epicardial potentials, i.e., the electrical potentials on the heart surface [49, 50]. The epicardial potentials provide an adequate representation of cardiac sources and are the only source model available for direct experimental validation, e.g., using multielectrode socks during open heart surgery. One benefit of this source model is the relatively low time and memory consumption because only the nodes on the heart surface are involved in the inverse computation. Another source model is developed for the reconstruction of the excitation propagation sequence on the heart surface [51, 52, 53]. It considers and simplifies the activation wavefront on the heart surface as a uniform dipolar double layer. The assumption of uniformity of the double layer makes the inverse problem less ill-posed. But it is not able to reflect the effects of unequal anisotropy ratios of cardiac tissue and it may lead to wrong results in case of infarction where the uniformity of the double dipole layer is not true. Because the relation between transmembrane voltages in the myocardium and the electrical potentials in the entire volume conductor is determined by the bidomain model, the transmembrane voltages are proposed as a source model for the inverse problem of ECG [54, 55]. The transmembrane voltages are defined in the whole active cardiac tissue. The reconstruction of transmembrane voltages images the real sources and gives a complete image of the heart condition.

In this thesis a new source model is proposed: the temporal integral of transmembrane voltages. The development of this source model is inspired by the uniform double layer (UDL) model. Using the UDL model the connection between the "activation time" during the cardiac cycle and ECG signal can be established [56]. It assumes that the transmembrane voltage at a source point behaves as a step function. Then, the original formulation of the inverse problem of electrocardiography can be rewritten as

$$A\tau = z_L \tag{5.9}$$

with

$$z_L = -\frac{\Delta t}{\Delta v} \sum_{\kappa=1}^{K} b(\kappa), \tag{5.10}$$

where τ stands for activation time, Δt denotes the step of temporal discretization, Δv is a constant value for the transmembrane voltage of an excited cardiac cell in the assumed step function, and κ indicates time index. This new formulation with activation time as source model remains linear. However, the transmembrane voltage cannot be treated as a step function and significant differences in action potential can be observed between different type of cells (see Fig. 2.4). Therefore, the problem becomes nonlinear when the assumption cannot hold. In this case a sophisticated optimization strategy must be employed in the inverse problem.

However, for some specific applications, e.g., in the identification of the origin of PVC only the first several *ms* after the beginning of excitation are needed to be considered, the development of an action potential in different type of cells can be considered to be approximately the same in this short period of time, e.g., in the first 20 *ms* (see Fig. 5.1). Thus, Eq. 5.2 remains linear if an integral is applied on both sides of Eq. 5.2 over the first several time steps as shown below.

$$A\mathbf{x} = \mathbf{b} \tag{5.11}$$

with

$$\mathbf{x} = \sum_{\kappa=K_1}^{K_2} x(\kappa) \tag{5.12}$$

$$\mathbf{b} = \sum_{\kappa=K_1}^{K_2} b(\kappa), \tag{5.13}$$

where $\mathbf{x}$ is the integral of transmembrane voltages and $\mathbf{b}$ the integral of ECG over the time interval between $\kappa = K_1$ and $\kappa = K_2$. It can be observed in Fig. 5.1 that the PVC origins indicated from the activation sequence and the integrals of transmembrane voltages are identical. This proves the validity of the new source model.

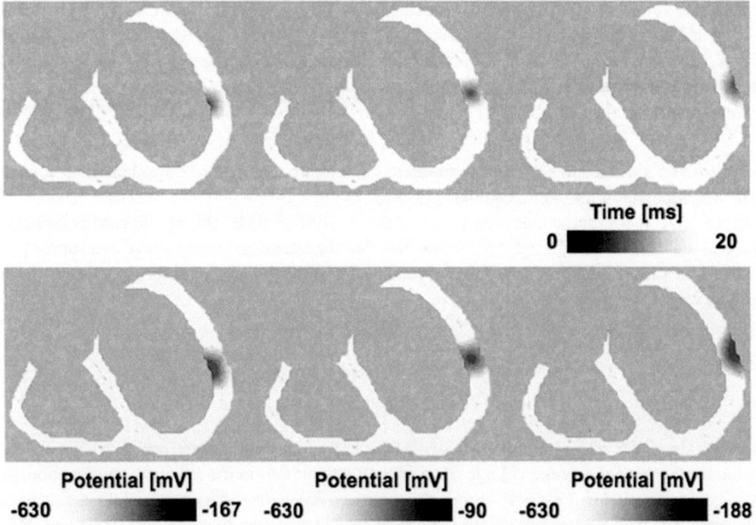

Fig. 5.1. Comparison between the activation time (the upper row) and the integral of transmembrane voltages (the bottom row). Three PVCs are shown in a cross-section of the ventricular model. The origin of PVC is indicated by the maximum in activation time and by the minimum in the integral of transmembrane voltages.

This model avoids the introduction of additional unknown parameters into the inverse problem like in the activation time formulation (Eq. 5.9 and Eq. 5.10), which makes the already ill-posed problem even more complicated. It still offers an image of the activation sequence in the entire heart tissue, from which the origin of PVC can be determined. The new source model amplifies the ECG signal recorded at the beginning of PVC, which is very weak. Applying the integral also reduces the effect of noise in the measurement.

In addition, this new source model can also be deployed in the reconstruction of myocardial infarction. In this application the integral is applied on the ST-segment, in which the transmembrane voltages are approximately constant both in the infarcted and healthy tissues (see Section 8.4).

5.4 Regularization Technique

Apparently, Eq. 5.2 can be solved in a direct way like

$$x = A^{-1}b, \tag{5.14}$$

or with the aid of the minimization problem below

$$x = \arg\min_{x}(\|b - Ax\|_2^2), \tag{5.15}$$

whose direct solution is given by

$$x = (A^T A)^{-1} A^T b, \tag{5.16}$$

where $\| \cdot \|_2^2$ denotes the $\ell 2$-norm.

However, due to the inherent ill-posedness of the inverse problem the columns or rows of the transfer matrix A are nearly linear dependent, i.e., it is not possible to compute the inversion of such an ill-conditioned matrix A as well as $A^T A$ in a straightforward way. Therefore, regularization techniques must be deployed in the solution of the inverse problem. Here three regularization methods from three different categories will be presented.

5.4.1 Tikhonov Regularization

Tikhonov regularization is the most widely used method to solve the inverse problem [57]. It is a representative of the penalty methods. Tikhonov regularization introduces a regularization term (penalty) in addition to the least squares residual:

$$x_\lambda = \arg\min_{x}(\|b - Ax\|_2^2 + \lambda^2 \|Lx\|_2^2) \tag{5.17}$$

where L denotes the regularization operator and λ is termed regularization parameter, which controls the weight attributed to the constraint condition $\|Lx\|_2^2$. In the case of the Tikhonov regularization 0-order regularization the regularization operator L is the identity matrix I. In the case of the 2nd order regularization L is a Laplacian operator.

The minimization problem (Eq. 5.17) is equivalent to the following set of augmented equations:

$$\begin{bmatrix} b \\ 0 \end{bmatrix} = \begin{bmatrix} A \\ \lambda L \end{bmatrix} [x_\lambda] \tag{5.18}$$

The solution to Eq. 5.18 is given by

$$x_\lambda = (A^T A + \lambda^2 L^T L)^{-1} A^T b. \tag{5.19}$$

5.4.2 TTLS Regularization

Truncated total least squares method (TTLS) is a regularization method based on the TLS formulation [58, 59]. It considers both the errors in the measurement signal b and in the transfer matrix A simultaneously. It solves the following minimization problem.

$$\min_{\tilde{A},\tilde{b}} \|(A,b) - (\tilde{A},\tilde{b})\|_F \quad \text{subject to } \tilde{b} = \tilde{A}x \tag{5.20}$$

where (A,b) is the augmented matrix with A and b side by side, $\|\cdot\|_F$ stands for the Frobenius norm, $\tilde{A}$ and $\tilde{b}$ are the contaminated A and b, respectively. The truncation (regularizaiton) is then performed on $(\tilde{A},\tilde{b})$ by applying singular value decomposition (SVD).

5.4.3 LSQR Regularization

Least squares QR-decomposition (LSQR) is an iterative regularization method in the category of Krylov subspace projection methods [60, 61]. LSQR iteratively computes three matrices using Lanczos bidiagonalization: an upper-bidiagonal matrix B_k and two matrices $U_k \equiv [u_1, \cdots, u_k]$ and $V_k \equiv [v_1, \cdots, v_k]$, with orthonormal columns. They satisfy the following conditions after k iterations:

$$b = \beta_1 u_1 = \beta_1 U_{k+1} e_1, \tag{5.21}$$

$$AV_k = U_{k+1} B_k, \tag{5.22}$$

$$A^T U_{k+1} = V_k B_k^T + \alpha_{k+1} v_{k+1} e_{k+1}^T, \tag{5.23}$$

where e_i denotes the ith unit vector.

Afterwards, the least squares problem is to be solved in the k-dimensional subspace $\mathscr{S}$, which is spanned by the first k vectors v_i:

$$\min_{x \in \mathscr{S}} (\|b - Ax\|_2^2). \tag{5.24}$$

A solution of the form $x^{(k)} = V_{(k)} y^{(k)}$ is sought in the k-dimensional subspace. Thus, the residual is defined as follows

$$r^{(k)} = b - Ax^{(k)} = \beta_1 u_1 - AV_k y^{(k)} = U_{k+1}(\beta_1 e_1 - B_k y^{(k)}). \tag{5.25}$$

This minimization problem is mathematically equivalent to the following formulation involving the bidiagonal matrix

$$y^{(k)} = \arg\min_{y^{(k)}} (\|\beta_1 e_1 - B_k y^{(k)}\|_2^2). \tag{5.26}$$

If the iteration stops after k steps, the solution is projected onto a k-dimensional subspace that gives a limited effect of regularization.

5.4.4 GMRes Regularization

Generalized minimal residual method (GMRes) is another iterative regularization method in the category of Krylov subspace projection methods [62, 63]. It is similar to the LSQR regularization, but it applies the Arnoldi iteration process to produce a B_k matrix (see Eq. 5.22) that is an upper Hessenberg matrix. It also requires that the transfer matrix A is a square matrix. Therefore, the transfer matrix is replaced by $A^T \cdot A$. In the iteration process the solution is sought, which minimizes the least squares problem in the jth Krylov subspace.

$$x_j = \arg \min_{x \in \mathcal{K}_j(A', y)} \|y - A' \cdot x\|, \tag{5.27}$$

where $\mathcal{K}_j$ denotes the j-th Krylov subspace, $A' = A^T \cdot A$ and $y = A^T \cdot b$.

5.4.5 LSQR-Tikhonov Hybrid Regularization

Iterative projection regularization methods like LSQR are able to transform the original problem onto a lower-dimensional space, but sometimes fail to provide sufficient regularization. Penalty methods like Tikhonov regularization are appropriate for constraining the inverse solution but they have to handle with problems of high dimension. The hybrid framework bridges these two kinds of regularization and takes the advantages of both of them [61, 64, 65, 66]. By combining LSQR regularization and Tikhonov regularization the following minimization problem is established in the k-dimensional Krylov subspace.

$$y_\lambda^{(k)} = \arg \min_{y^{(k)}} (\|\beta_1 e_1 - B_k y^{(k)}\|_2^2 + \lambda^2 \|L y^{(k)}\|_2^2). \tag{5.28}$$

5.4.6 Maximum *a posteriori* Based Regularization

The maximum *a posteriori* (MAP) based regularization belongs to the group of Bayes estimators. Its mathematical theory was first taken out in 1960's by applying the idea of the famous Wiener filter to matrix inversion [67]. The MAP-based regularization is capable of the direct incorporation of *a priori* statistical information about the unknown cardiac sources into the solution of inverse problem. From 1970's it has been applied in the inverse problem in different areas, e.g., in the inverse problem of ECG [68, 69] and in the inverse problem in geophysics [70].

In this case errors are included in the formulation of the inverse problem:

$$b = Ax + e, \tag{5.29}$$

with the assumption that the means of x and e are zero.

A matrix H is introduced as an optimal estimation of A^{-1}. Thus, the corresponding approximate solution $\hat{x}$ to the inverse problem is

$$\hat{x} = Hb. \tag{5.30}$$

Combining the Eq. 5.29 and Eq. 5.30 it gives

$$\hat{x} = HAx + He. \tag{5.31}$$

The "estimation error" is defined as the difference between the approximated solution $\hat{x}$ and the exact solution x

$$\hat{x} - x = (HA - I)x + He, \tag{5.32}$$

where the first term refers to the resolving error caused by the inaccuracy of H and the second term refers to the random errors included in the measurement. The smaller both terms are, the closer is the approximate solution $\hat{x}$ to the exact solution x. The covariance matrix of the estimation error is given by

$$C = (HA - I)C_x(HA - I)^T + HC_eH^T, \tag{5.33}$$

where C_x and C_e are the covariance matrices of the approximate solution and errors in the measurement, respectively. Similarly, smaller covariance of the estimation error leads to better solution $\hat{x}$. The "minimum variance" estimator

$$H = C_xA^T(AC_xA^T + C_e)^{-1} \tag{5.34}$$

is derived to minimize the covariance of estimation errors C regardless of the form of the *a priori* probability density for x [70].

In this way, the approximate solution $\hat{x}$ of the inverse problem is obtained:

$$\hat{x} = C_xA^T(AC_xA^T + C_e)^{-1}b \tag{5.35}$$

Because the estimator H maximizes (with respect to x) the *a posteriori* probability of the observed data b conditional to the solution x, this method is called maximum *a posteriori* based regularization [71].

In the MAP-based regularization C_x is a statistical description of the the cardiac sources. It can be estimated from experience or be directly extracted from the simulation results. When a stochastical basis $\mathscr{X}$ including N possibilities (observations) of cardiac sources at the time instant of interest is available, i.e., $\mathscr{X} = [x_1 \; x_2 \; \cdots \; x_N]$. The covariance matrix C_x is calculated as follows.

$$C_x = \frac{1}{N}(\mathscr{X} - \bar{\mathscr{X}})(\mathscr{X} - \bar{\mathscr{X}})^T, \tag{5.36}$$

where N is the number of observations included in $\mathscr{X}$ and $\bar{\mathscr{X}}$ denotes the mean vector of $\mathscr{X}$.

5.4.7 Regularization Parameter Choice Method

The determination of the optimal value for the regularization parameter is critical for the quality of the inverse solution. Parameter choice method can help to automatically determine the optimal value for regularization parameter. Two parameter choice methods, i.e., the "L-curve" method and Generalized Cross Validation (GCV), are utilized in the present thesis.

"L-curve" is a widely used parameter choice method for the inverse problem based on heuristic observations [72, 73]. When Tikhonov regularization is applied, this method plots the following functional for a wide range of different values of the regularization parameter λ.

$$\|Lx_\lambda\|_2^2 = f(\|Ax_\lambda - b\|_2^2) \tag{5.37}$$

Often, the plotted curve has a shape of an "L". The optimal value of λ is found at the point of the maximal curvature of the L-curve, where the balance point between the least squares residual and the regularization term is found.

Generalized cross validation (GCV) is another parameter choice method, which is based on statistical considerations [74]. It minimizes the following function.

$$G = \frac{\|Ax_\lambda - b\|_2^2}{(Tr(I - AA_\lambda^\dagger))^2} \tag{5.38}$$

where $A_\lambda^\dagger$ is the so-called inverse regularized operator of the transfer matrix A, which comes with the regularization method applied. For Tikhonov regularization $A_\lambda^\dagger = (A^T A + \lambda^2 I)^{-1} A^T$.

The L-curve method is applicable to the penalty methods like Tikhonov regularization. The GCV method can also be utilized in other regularization methods, e.g., in TTLS to determine the optimal truncation parameter and in the MAP-based regularization to select the appropriate value for the error estimation C_e. In the hybrid regularization methods like the LSQR-Tikhonov regularization, both parameter choice methods can be employed in the internal regularization process.

5.4.8 Spatio-Temporal Approach

Because both ECG signal and cardiac sources are temporally correlated, the use of temporal information brings additional advantage compared to solving the problem frame-by-frame. For this reason, several spatio-temporal approaches are developed for the inverse problem of ECG [75].

A straightforward way to incorporate the temporal information is to place the cardiac sources and the measurement data at different time instants in a long vector as follows:

$$\mathbf{x} = \left(x_1{}^T, x_2{}^T, \cdots x_n{}^T\right)^T \tag{5.39}$$

$$\mathbf{b} = \left(b_1{}^T, b_2{}^T, \cdots b_n{}^T\right)^T. \tag{5.40}$$

where n is the number of time instants.

In this case the transfer matrix becomes a diagonal block matrix constructed by the transfer matrix applied in the spatial version

$$\mathbf{A} = \begin{bmatrix} A & 0 & \cdots & 0 \\ 0 & A & & \vdots \\ \vdots & & \ddots & 0 \\ 0 & \cdots & 0 & A \end{bmatrix}. \tag{5.41}$$

The spatio-temporal form of the inverse problem is

$$\mathbf{Ax} = \mathbf{b}, \tag{5.42}$$

However, the consumption of computational time and memory grows dramatically as the number of time instants involved in the problem increases when this approach is deployed.

Another spatio-temporal approach is proposed by Greensite [76]. In Greensite's spatio-temporal framework the source matrix and the measurement matrix are constructed in the following way:

$$X = \left(x_1, x_2, \cdots x_n\right) \tag{5.43}$$

$$B = \left(b_1, b_2, \cdots b_n\right). \tag{5.44}$$

The inverse problem then has the form

$$AX = B, \tag{5.45}$$

where A is the same transfer matrix as that in the spatial version.

First, the singular value decomposition (SVD) is performed on the measurement matrix B.

$$B = U\Sigma V^*, \tag{5.46}$$

where U and V are unitary matrices containing the left and right singular vectors, respectively, and Σ is a diagonal matrix with nonnegative singular values on the diagonal.

Then, the inverse problem is mapped onto the orthogonal space spanned by the temporal (right) singular vectors of B as

$$AQ = Y \tag{5.47}$$

with $Q = XV$, $Y = BV = U\Sigma$.

Afterwards, the projected problem is solved "spatially" using a regularization method of your choice, i.e., each column of Y is considered separately. According to [77] only the first column vectors in Y, which satisfy the discrete Picard condition, are included in the inverse calculation. The rest part of Y is truncated and excluded in the calculation.

At the end, the inverse solution $\hat{Q}$ is transformed back to the original space by multiplying V^* on the right side.

$$\hat{X} = \hat{Q}V^*. \tag{5.48}$$

Thus, the final solution $\hat{X}$ in the time period under consideration is achieved.

5.4.9 Spatio-Temporal Maximum *a posteriori* Based Regularization

By applying the Greensite's spatio-temporal framework to the MAP-based regularization, the problem (Eq. 5.47) is solved "spatially" in the orthogonal temporal space as follows.

$$\hat{q} = C_q A^T \left(A C_q A^T + C_\varepsilon\right)^{-1} y, \tag{5.49}$$

where q is one column vector in Q and y is the corresponding column vector in Y, and ε is the errors e after being projected onto the orthogonal temporal space.

For the spatio-temporal MAP-based regularization the calculation of the covariance matrix C_q is also performed in the orthogonal temporal space: First, the N possibilities (observations) of cardiac sources in the time period of interest are projected onto the orthogonal temporal space by multiplying the right singular vector V of the ECG signal B applied to the inverse problem.

Then, the projected cardiac sources Q are treated "spatially". The ith column vectors of all the N observations are stored in a matrix $\mathscr{Q} = [q_1 \ q_2 \ \cdots \ q_N]$ and the covariance matrix for this column vector is computed as shown below.

$$C_q = \frac{1}{N}(\mathscr{Q} - \bar{\mathscr{Q}})(\mathscr{Q} - \bar{\mathscr{Q}})^T, \tag{5.50}$$

where N is the number of observations included in $\mathscr{Q}$ and $\bar{\mathscr{Q}}$ denotes the mean vector of $\mathscr{Q}$.

5.4.10 Spatio-Temporal LSQR-Tikhonov Hybrid Regularization

The spatio-temporal LSQR-Tikhonov hybrid regularization combines the spatio-temporal regularization framework and the hybrid regularization framework [78]. First, the inverse problem is mapped onto the orthogonal temporal space by applying Greensite's appraoch. The problem is then further projected onto the k-dimensional Krylov subspace using LSQR iteration method. Afterwards, the inverse problem is solved using Tikhonov regularization in the Krylov subspace. At the end, the inverse solution is transformed back to the original space. In Fig. 5.2 The process of solving the inverse problem of ECG using the spatio-temporal LSQR-Tikhonov hybrid method is illustrated.

5.5 Optimization of Electrode Positions for a BSPM System

The objective of this study to find an optimal electrode configuration of the body surface mapping system for the reconstruction of myocardial infarction using the inverse problem of ECG. The inverse problem of ECG is normally strongly underdetermined. The major cause of the underdetermination is the limited spatial resolution of ECG measurement, i.e., the number of measurement electrodes is insufficient or not well arranged to cover the important signal on the body surface. The quality of the inverse solutions can be improved by increasing the electrode quantity. Whereas, an ECG measurement system consisting of too many recording electrodes will be impractical in the clinical practice and some of the electrodes can be redundant. Therefore, the electrode positions of the BSPM system should be optimized so that the most important and useful parts of body surface signals can be recorded with an adequate spatial resolution. Many investigations have been undertaken in the optimization of the electrode positions for the BSPM system regarding different criterions, e.g., the spatial frequency analysis [79], the null-space theory [80] and the local linear dependency (LLD) maps [81]. However, these investigations are done for the general application of the inverse problem of ECG. In the current study the optimization concentrates on a specific application: the reconstruction of myocardial infarction.

Because the most significant symptom of myocardial infarction in ECG is ST-elevation or ST-depression, the optimal electrode configuration is reached when it is able to cover the most significant spatial patterns in BSPMs caused by myocardial infarction during the ST-segment. This study is performed on the anatomical model of Patient 1 (see Fig. 3.2) and a 64-channel BSPM system is to be optimized. A data base (stochastical basis) including 153 different myocardial infarctions in the entire left ventricular wall and the ventricular septum is created. The cardiac activities of myocardial infarction are simulated using the cellular automaton and the corresponding body surface potentials are calculated using the bidomain model and the finite element method. The ST-integral maps of these simulated BSPMs (see Eq. 4.23) are analyzed with the aid of the

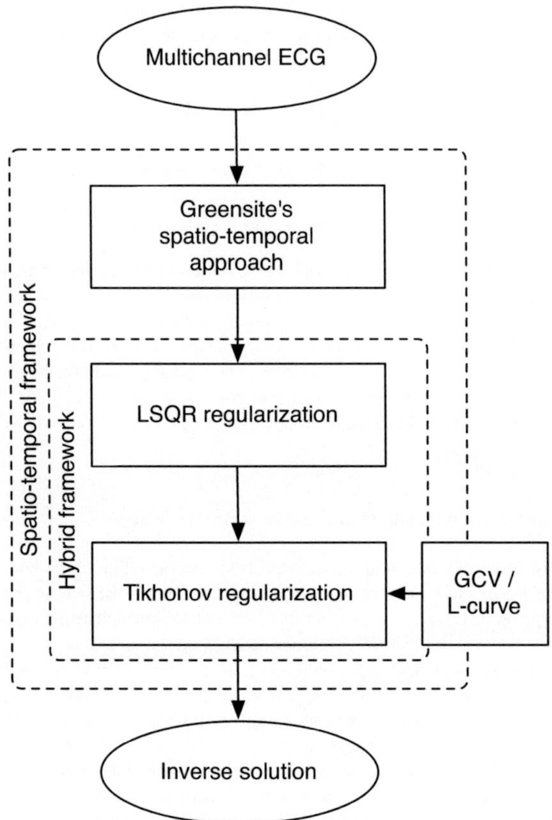

Fig. 5.2. Flow diagram demonstrating the process of solving the inverse problem of ECG by combining the spatio-temporal and the LSQR-Tikhonov hybrid regularization frameworks [46].

singular value decomposition (SVD).

The ST-integral maps of all 153 myocardial infarctions in the stochatical basis are stored in a matrix $\mathbf{B} = (b_1, b_2, \cdots b_{153})$. The SVD is performed on the matrix $\mathbf{B}$:

$$\mathbf{B} = U\Sigma V^* = \sum_{i=1}^{n} u_i s_i v_i^T, \tag{5.51}$$

where U is a unitary matrix containing left singular vectors, Σ is a diagonal matrix with nonnegative singular values on the diagonal, V is a unitary matrix containing right singular vectors, u_i is the ith left singular vector, s_i is the ith singular value and v_i is the ith right singular vector.

The left singular vectors u_i of the matrix $\mathbf{B}$ are the spatial patterns that occur in the body surface potentials during ST-segment. Every ST-integral map can be considered as a linear combination of these spatial patterns. Because the singular values s_i are sorted in a descending order, the first several left singular vectors with large singular values are more influential than those with small singular values. In the present study the singular values show a dramatic decrease after the 5th singular value. Hence, the first 5 left singular vectors are taken into account and the rest of them are negligible in the analysis. The optimal electrode configuration of the BSPM system should cover the regions on the body surface, where the first 5 left singular vectors show the strongest magnitudes.

6

Intracardiac Mapping and Navigation

6.1 Introduction

Nowadays, catheter interventions are widely used in the treatment of cardiac arrhythmias, e.g., radiofrequency catheter ablation of atrial flutter [82, 83, 84, 85, 86]. Intracardiac mapping and navigation are an essential prerequisite for the catheter interventional operations. In the conventional mapping and navigation X-ray fluoroscopy using a mono- or bi-plane system is deployed. During the interventional procedure the mapping and navigation data have to be registrated with the off-line geometrical information, which is normally acquired using CT or MRI beforehand. Moreover, the intracardiac mapping and navigation based on the fluoroscopy bear drawbacks like low resolution, inaccuracy and X-ray exposure [87, 88, 89]. In the late 1990's revolutionary techniques have been developed, e.g., the Biosense Carto system (see Fig. 6.1 a) and the St. Jude Medical NavX 3D Mapping System (see Fig. 6.1 b), which allow real-time 3D *in vivo* cardiac mapping and navigation with high resolution and accuracy [83, 86, 90, 91, 92, 93, 94, 95]. Such systems are also able to reconstruct the chamber geometry and generate a detailed 3D map of cardiac activities during the interventional procedure [87, 88, 92, 96, 97, 98, 99].

Because the intracardiac mapping and measurement with only one catheter electrode could be very time-consuming, it is suggested to introduce multiple electrodes into the system. In this way, a plurality of sites on the endocardium can be sampled and measured simultaneously and accordingly the performance of the cardiac mapping and navigation system can be improved significantly. However, it can lead to high cost when every catheter electrode is equipped with a localizer. Furthermore, more than one type of catheter and even catheters from different companies have to be used simultaneously in cardiac interventional procedures. Therefore, it is difficult to find a solution to localize all these catheter electrodes efficiently and with a sufficient accuracy.

In the present thesis an impedance based catheter positioning system is proposed, which enables the simultaneous localization of multiple electrodes on a catheter with high accuracy. Furthermore, no expensive specialized hardware is required in addition to the currently existing mapping and navigation system. The localization of catheters of different kinds and from different companies is also allowed in the new system.

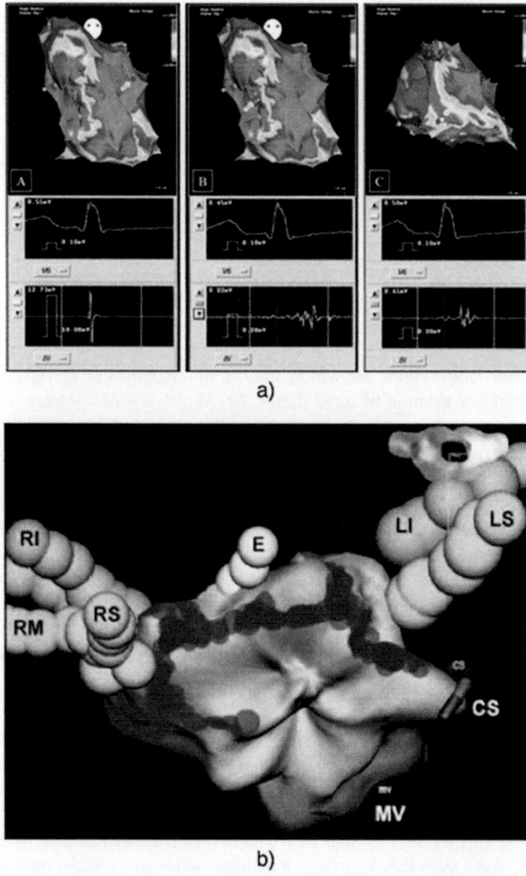

Fig. 6.1. An abnormal electroanatomic voltage map of the right ventricle measured using Carto XP system shown in 3 views (the upper panels in a) and electrograms recorded at several sites on the endocardium (the bottom panels in a) [83], and a NavX map of the left atrium and accompanying pulmonary veins established during ablative intervention for atrial fibrillation (b) [86].

6.2 Impedance Based Catheter Positioning System

The development of the impedance based catheter positioning system is supported by computer simulation. The current simulation study is performed on the male Visible Human model (see Fig. 3.1) using the finite element method. 6 cylindrical electrode-patches attached on the body surface are introduced to measure the currrents between the catheter electrode and the patches as shown in Fig. 6.2. 3 patches are on the front of the thorax and the other 3 are on the back.

The patches are metallic and all of them have the same dimension with a radius of 4 *cm* and the height of 2 *mm*. The gap between the patch and the skin is filled with a high-conductivity gel. The catheter electrode in the heart has a radius of 1 *mm*. In order to reduce the numerical error in the computer simulation, an intelligent refinement of the mesh is performed, i.e., a simulation is run on the initial model, and then all the elements in the mesh with high potential gradients are subdivided [6]. Additional mesh refinement is also done in the region around the patches to increase the accuracy of the measurement of currents in the simulation.

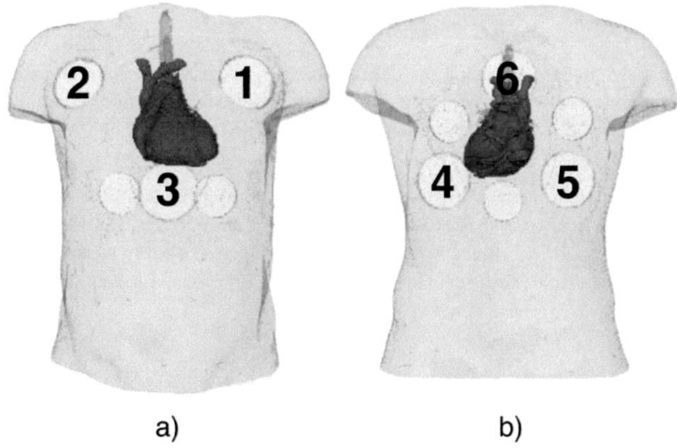

Fig. 6.2. 6 patches attached on the torso surface for impedance measurement: anterior view (a) and posterior view (b). The torso is displayed half-transparently and the position of the heart is shown. The patches are numbered from 1 to 6 [100].

During the mapping and navigation procedure several catheters or a catheter with multiple electrodes are inserted into one chamber of the heart. One of these electrodes is localisable, whose location can be determined with high precision, e.g., using the magnetic localizer. The other electrodes are non-localisable. In addition, an AC or DC voltage (1 *V* in the simulation study) is applied between the catheter electrode under consideration and the patches on the body surface in the localization. The patches are connected to the ground (0 *V*) during the simulation. The currents that flow from the catheter electrode to all 6 patches are measured simultaneously. Then, the normalized currents at the patches are calculated.

$$\bar{I}_i = \frac{I_i}{(\sum_{i=1}^{6} I_i)/6},$$

(6.1)

where i_i denotes the current measured at patch i and $\bar{I}_i$ indicates the normalized current obtained at patch i. The rationale to apply the normalized currents is as follows: the human body is a strongly inhomogeneous environment, hence the relationship between the currents (not normalized) and the electrode position is not linear. Whereas, a quasi-linear relationship between the normalized currents and the electrode position can be observed. This method also minimizes other problems that rise up in clinical practice, e.g., catheter electrodes may be of various sizes and different shapes. The currents through the surface patches change proportionally with the change of electrode size, but no difference shows after the normalization. The variation of electrode shapes also has only minimal influence on the normalized currents. Moreover, the currents measured at patches are directly proportional to the voltage applied at the catheter electrode, the amount of voltage has no effect on the normalized currents. Please note that this method is developed and patented by Biosense Webster.

As aforementioned, the method proposed in the current study is based on the quasi-linear relationship between the normalized currents from the catheter electrode to the body surface patches and the electrode position.

$$\bar{I} = A \cdot P, \tag{6.2}$$

where $\bar{I}$ is a $6 \times n$ matrix containing the normalized currents at the 6 patches for n electrode positions, P is a $3 \times n$ matrix containing the x, y, z coordinates for n electrode positions, and A is a 6×3 transfer matrix describing the quasi-linear relationship between $\bar{I}$ and P.

Before the localization begins, the system is calibrated using the localisable catheter, i.e., the catheter is navigated to several sites and the normalized currents $\bar{I}$ at all patches are recorded for every electrode position. At the same time, the coordinates of every electrode position P are delivered by the localisable catheter. Thus, the transfer matrix A can be constructed.

$$A = \bar{I} \cdot P^{\dagger}, \tag{6.3}$$

where $P^{\dagger}$ is the Moore-Penrose pseudo-inverse of P.

Once the quasi-linear system A is established, the measurement is performed on the other non-localisable electrodes. The locations of these electrodes are then determined from the corresponding normalized currents $\bar{I}$ obtained at the patches as follows.

$$\tilde{P} = A^{-1} \cdot \bar{I}. \tag{6.4}$$

In Eq. 6.4 the inverse of A can be derived from Eq. 6.3

$$A^{-1} = P \cdot \bar{I}^{\dagger}, \tag{6.5}$$

where $\bar{I}^{\dagger}$ is the Moore-Penrose pseudo-inverse of $\bar{I}$.

Part III

Results

Results: Cardiac Modeling and Forward ECG Simulation

7.1 Realistic Environment

In order to investigate the influence of different kinds of error on the solution of the inverse problem of ECG a realistic environment is created. The following errors in the measurement and in the model are considered: 30 *dB* measurement noise, 0.1 *Hz* baseline wander, 5 *mm* inaccuracy in the localization of body surface electrodes, 20% inaccuracy in the estimation of tissue conductivities and the modeling errors introduced by neglecting the heart motion and respiration.

In the realistic environment as described in Section 4.4 a respiratory cycle with a length of 4 *s* including 4 cardiac cycles of the sinus rhythm is simulated. The simulated ECG is saved every 4 *ms*. In addition, simulations including only one type of error at each time are also performed. Thus, the effect of every type of error both on ECG and on the inverse solution can be observed separately. An ECG simulated in an idealistic environment (without any error applied) serves as reference. The simulation results are presented in Fig. 7.2 and Fig. 7.3. ECG channels 15 and 45 are shown for each case. The electrode position associated with these two channels are marked in Fig. 7.1. When no lung motion is involved in the simulation, only one cardiac cycle is shown.

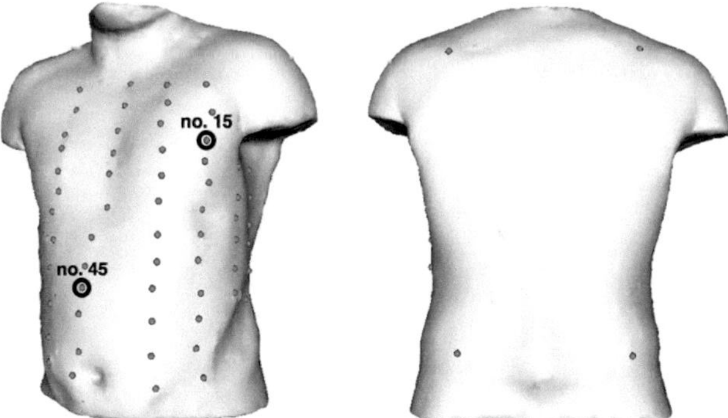

Fig. 7.1. Electrode configuration of the 64-channel ECG shown on the front (left) and on the back (right) of the Visible Human torso model.

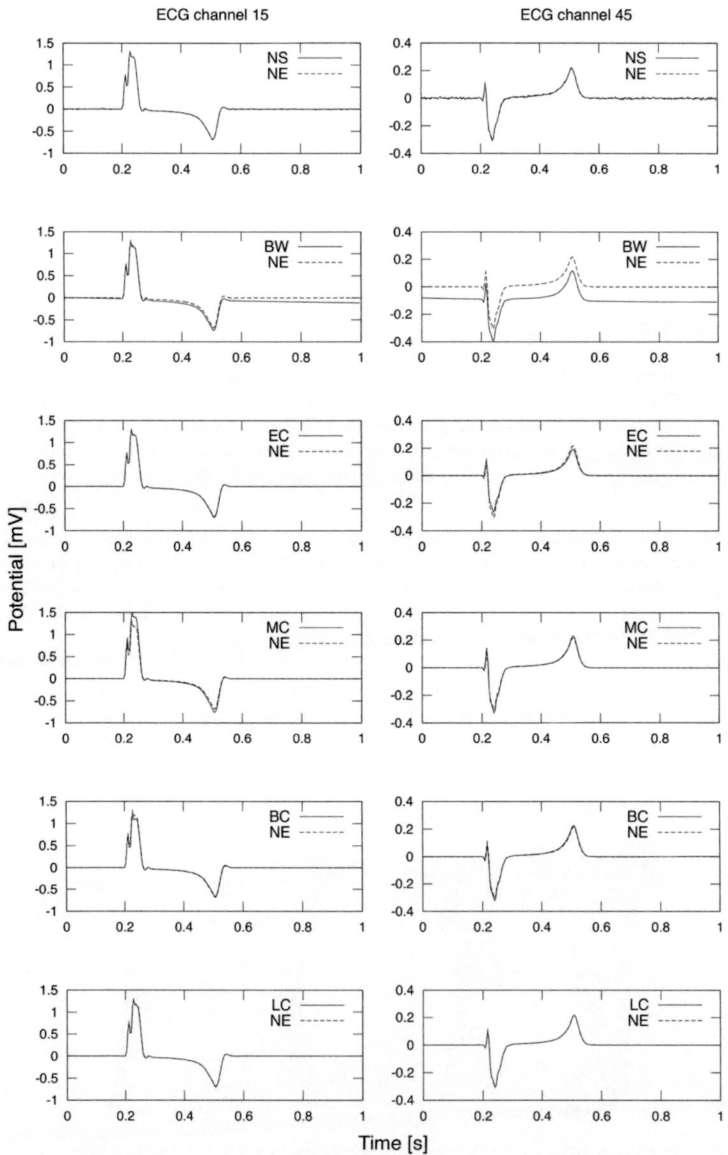

Fig. 7.2. Influence of different kinds of error on ECG signal. NE: no error, NS: 30 dB noise, BW: 0.1 Hz baseline wander, EC: 5 mm localization of electrodes, MC: 80% conductivity of myocardium, BC: 80% conductivity of blood inside the heart chambers, LC: 80% conductivity of lungs [46].

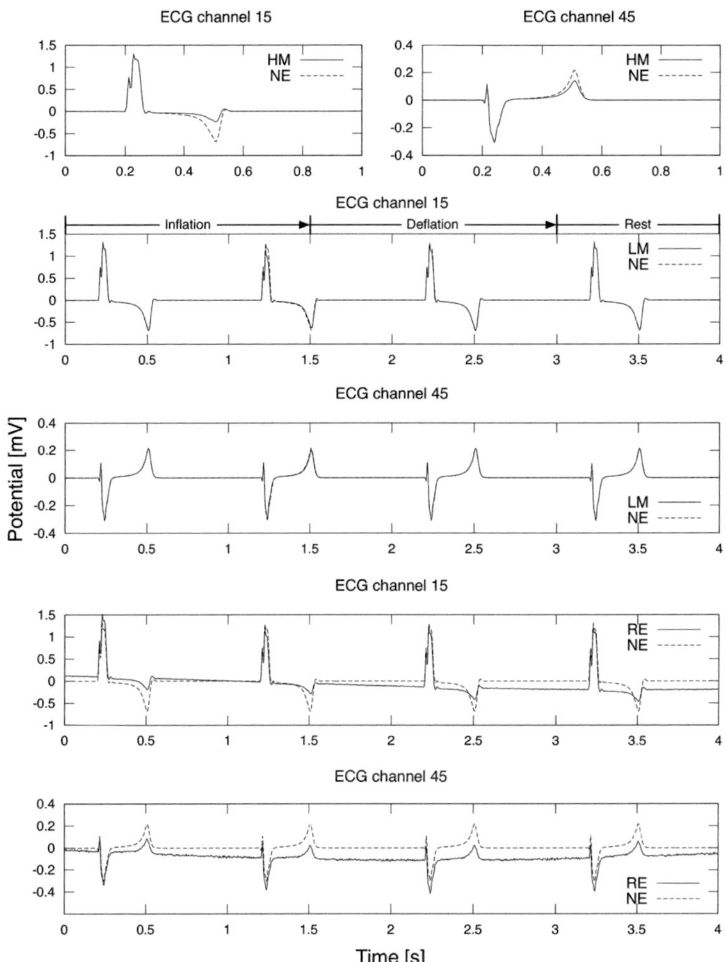

Fig. 7.3. Influence of different kinds of error on ECG signal (the first row to the third row) and the ECG signal in the realistic environment (the fourth and fifth rows). NE: no error, HM: heart motion, LM: lung motion, RE: realistic environment [46].

When the lung motion is involved, a full respiratory cycle including 4 cardiac cycle is shown.

Discussion

As can be seen in Fig. 7.2 and Fig. 7.3 the 20% decrease of the original conductivities in blood and lungs, and the 5 *mm* shifting of electrodes do not result in significant changes in ECG. The respiration only causes minor changes in the ECG amplitude during the second cardiac cycle, where the lung volume is at its maximum and the conductivity of lungs reaches its minimum. The 20% change of the conductivity in myocardium leads to a remarkable increment of the amplitude of R-peak. The baseline wander adds a significant oscillation to the ECG signal. The heart motion leads to remarkable changes in the morphology and amplitude of T-wave.

7.2 Premature Ventricular Contraction

In the current simulation, 243 PVCs are simulated in the entire left ventricle. The simulation results will be applied as *a priori* information in the MAP-based regularization (see Section 5.4.6) for the localization of the origin of PVC. The simulations are performed on Patient 2 (see Fig. 3.3) using the cellular automaton. For a complete coverage of possible PVCs originating from various sites in the left ventricle, the left ventricular model is subdivided into 81 segments as shown in Fig. 7.4. In each segment 3 PVCs are simulated, with the ectopic center on the endocardium, in the middle of the ventricular wall and on the epicardium. Thus, in total 243 (81 × 3) PVCs are simulated in the entire left ventricle including the septum. Moreover, the forward ECG simulation is also performed.

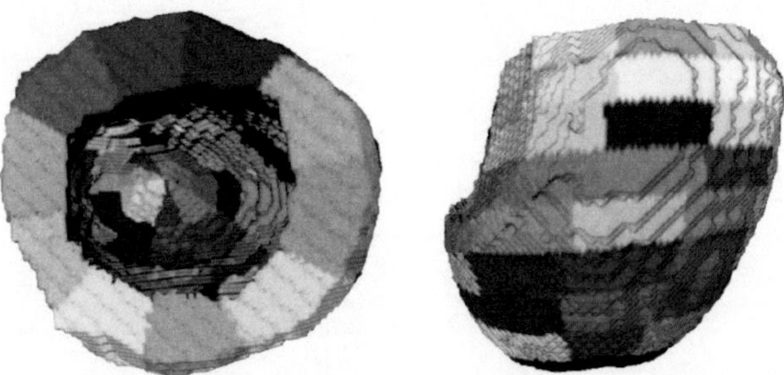

Fig. 7.4. Subdivision of the left ventricle and the septum into 81 segments

In Fig. 7.5 three examples out of the 243 simulated PVCs are shown. These three PVCs are originated from the same apical lateral segment, but starting from endocardium, midmyocardium and epicardium. In the upper row of Fig. 7.5 the integrals of transmembrane voltages over the time interval from 8 *ms* before to 20 *ms* after the beginning of PVC are shown on a cross-section of the ventricular model. The origin of PVC is identified by the minimum in the integral of transmembrane voltages (see Fig. 5.1). In the bottom row of Fig. 7.5 the integrals of body surface potentials associated with the three PVCs during the same time interval are displayed.

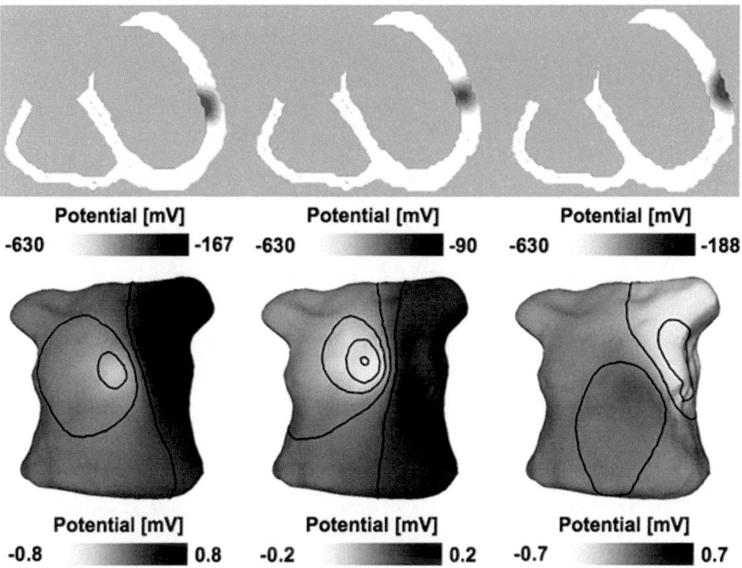

Fig. 7.5. The transmembrane voltage integrals over 8 *ms* before and 20 *ms* after the beginning of PVCs (the upper row) and the body surface potential integrals over 8 *ms* before and 20 *ms* after the beginning of PVCs (the bottom row). The PVCs are initiated from endocardium (the left column), midmyocardium (the middle column) and epicardium (the right column).

Discussion

From the integrals of body surface potentials in Fig. 7.5 it can be observed that the patterns appearing on the body surface are remarkably different for the PVCs initiated from endocardium and epicardium. Whereas, the difference between the body surface potentials generated by the PVCs starting from endocardium and midmyocardium is only shown in the potential amplitude.

7.3 Myocardial Infarction

In the current thesis, a meaningful and comprehensive stochastical basis including various myocardial infarctions at different sites and of different sizes in the left ventricle is created for each of the two personalized anatomical models (see Patient 1 and Patient 2 in Section 3.1). For this purpose, the left ventricular model is subdivided into 17 segments according to the recommendation of the American Heart Association for the segmentation of left ventricle (see Fig. 7.6)

[101, 102]. The subdivision of the two models are illistrated in Fig. 7.7. In each segment 3 types of infarction are simulated, i.e., subendocardial, transmural and subepicardial infarctions. For each type, 3 different sizes are considered, i.e., 10 *mm*, 20 *mm* and 30 *mm* for the radius of infarction. Thus, a stochastical basis composed of 153 ($17 \times 3 \times 3$) different simulated myocardial infarctions in the whole left ventricle and the ventricular septum is established.

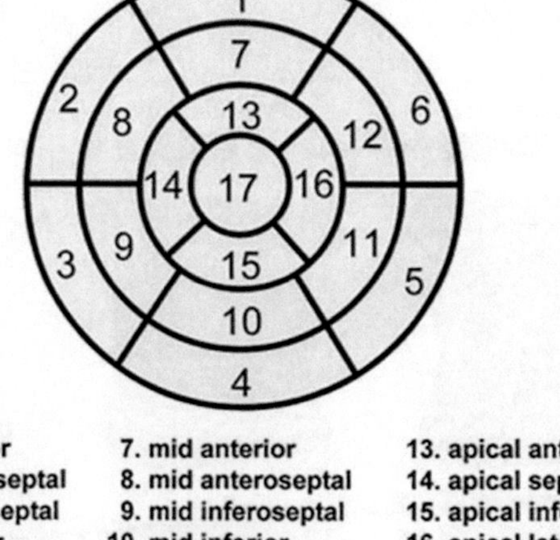

1. basal anterior	7. mid anterior	13. apical anterior
2. basal anteroseptal	8. mid anteroseptal	14. apical septal
3. basal inferoseptal	9. mid inferoseptal	15. apical inferior
4. basal inferior	10. mid inferior	16. apical lateral
5. basal inferolateral	11. mid inferolateral	17. apex
6. basal anterolateral	12. mid anterolateral	

Fig. 7.6. The nomenclature of 17 AHA segments in left ventricle [101].

The stochastical basis is deployed in three investigations in terms of the detection and localization of myocardial infarctions: the optimization of the electrode positions for a wearable ECG monitoring system (see Section 7.4), the reconstruction of myocardial infarctions by solving the inverse problem using the MAP-based regularization (see Section 8.3) and the optimization of the electrode positions for a BSPM system (see Section 8.4).

The simulation of myocardial infarctions is conducted on the two personalized models using the cellular automaton. Forward calculation of ECG is performed for all simulated myocardial infarctions in the stochastical basis and the corresponding BSPMs are obtained. As an example, 6 myocardial infarctions simulated on the model of Patient 1 are displayed in Fig. 7.8. In the upper row of Fig. 7.8 the 3 myocardial infarctions are in the apical lateral segment and of 3 different sizes. In the bottom row of Fig. 7.8 the 3 apical myocardial infarctions have the same size but different depths in the ventricular myocardium. The BSPMs associated with these 6 myocardial infarctions are shown in Fig. 7.9. Furthermore, the corresponding ST-integral maps are shown in Fig. 7.10.

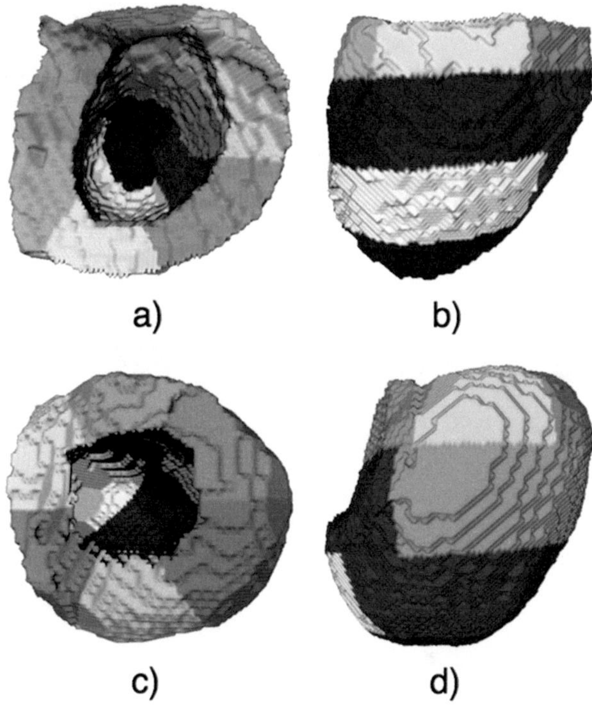

Fig. 7.7. The subdivision of the left ventricular models of Patient 1 (a and b) and of Patient 2 (c and d) into 17 segments according to the AHA recommendation. The models are shown in two aspects, respectively.

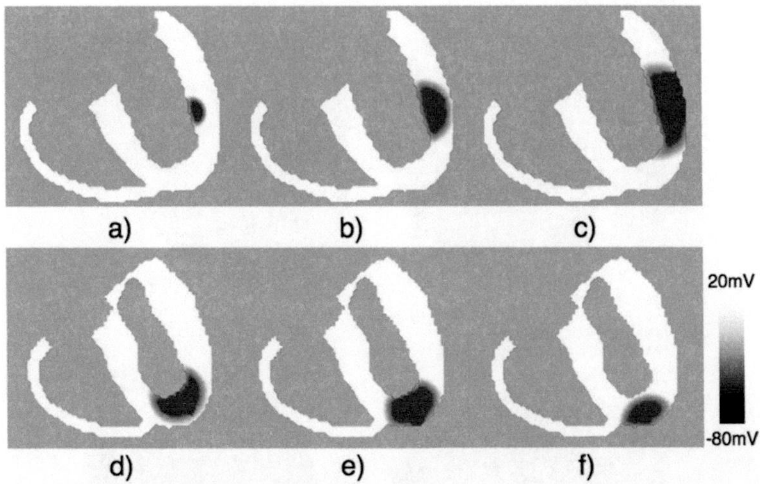

Fig. 7.8. Simulated apical lateral infarctions (Segment 16) with radii of 10 *mm* (a), 20 *mm* (b) and 30 *mm* (c); simulated subendocardial (d), transmural (e) and subepicardial (f) infarctions with a radius of 20 *mm* in the apical segment (Segment 17). The simulations are performed on the model of Patient 1. The distribution of transmembrane voltages in the middle of ST-segment is shown in a cross-section of the ventricular model [103].

Discussion

By comparing Fig. 7.9 and Fig. 7.10 it can be seen that the BSPMs at the middle of ST-segment and the integrals of BSPMs during ST-segment show similar patterns, because the transmembrane voltages both in the infarcted tissue and in the healthy tissue stay approximately constant during ST-segment. The electrical potentials on the body surface show the same behavior.

As shown in Fig. 7.9 a to c and Fig. 7.10 a to c, the amplitude of body surface potentials is proportional to the size of myocardial infarction and the patterns of BSPM remain roughly unchanged. By considering Fig. 7.9 d to f and Fig. 7.10 d to f it can be observed that the subendocardial, transmural and subepicardial infarctions can be differentiated from the patterns appearing in BSPM. Moreover, the subendocardial infarction generates an approximately reversed pattern in BSPM as the subepicardial infarction (see Fig. 7.9 d and f and Fig. 7.10 d and f).

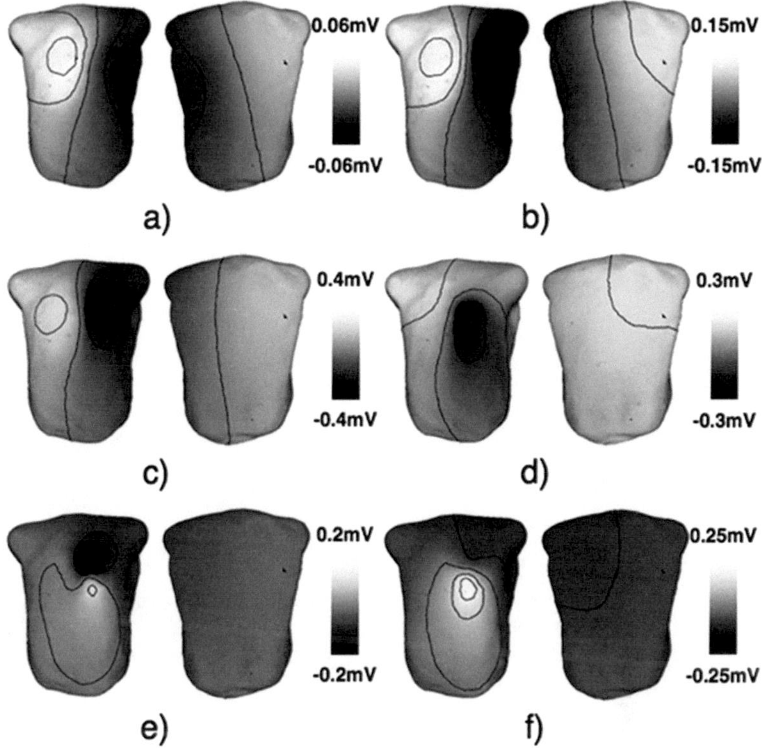

Fig. 7.9. The BSPMs corresponding to the simulated apical lateral infarctions (Segment 16) with radii of 10 *mm* (a), 20 *mm* (b) and 30 *mm* (c); the BSPMs corresponding to the simulated subendocardial (d), transmural (e) and subepicardial (f) infarctions with a radius of 20 *mm* in the apical segment (Segment 17). The simulations are performed on the model of Patient 1. The BSPMs in the middle of ST-segment are shown [103].

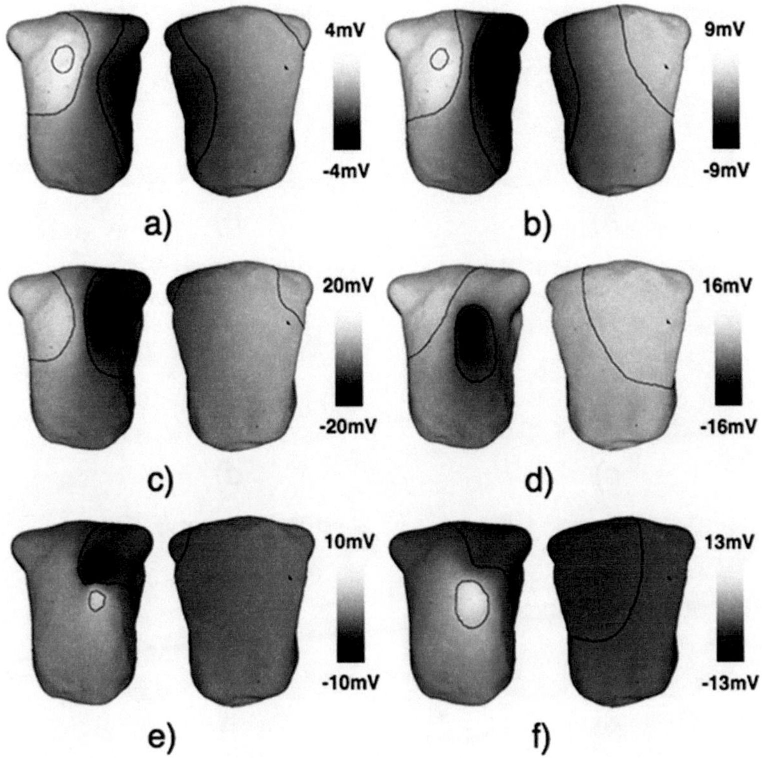

Fig. 7.10. The ST-integral maps corresponding to the simulated apical lateral infarctions (Segment 16) with radii of 10 *mm* (a), 20 *mm* (b) and 30 *mm* (c); the ST-integral maps corresponding to the simulated subendocardial (d), transmural (e) and subepicardial (f) infarctions with a radius of 20 *mm* in the apical segment (Segment 17). The simulations are performed on the model of Patient 1 [47].

7.4 Optimization of Electrode Positions for a Wearable ECG Monitoring System

This simulation study is aimed to determine an optimal electrode configuration of a wearable
ECG monitoring system with a minimal number of electrodes for the detection of myocardial
infarctions. The optimization strategy is described in Section 4.5. The study is performed on two
anatomical models (Patient 1 and Patient 2). As the initial value the threshold T for the difference
between $\mathbf{D}_k(i,j)$ and $\mathbf{D}_h(i,j)$ is set to 2 mV. During the optimization process the threshold T decreases with a step of 0.5 mV. Until the threshold T is set to 0.1 mV, still no satisfying result is
obtained on both models, i.e., a single pair of electrodes is not able to detect all 153 myocardial
infarctions of the stochastical basis. With 2 pairs of electrodes with one common electrode (3
electrodes) an optimal electrode configuration can be first determined at a threshold of 0.4 mV for
Patient 1 and 0.35 mV for Patient 2.

The optimal electrode configurations obtained on the model of Patient 1 and on the model of
Patient 2 are shown in Fig. 7.11. In the optimal electrode configurations one electrode is located
below the fossa jugularis and slightly to the right of the thorax, and one electrode is on the left
back for both patients. For Patient 1 the third electrode is located on the left chest directly above

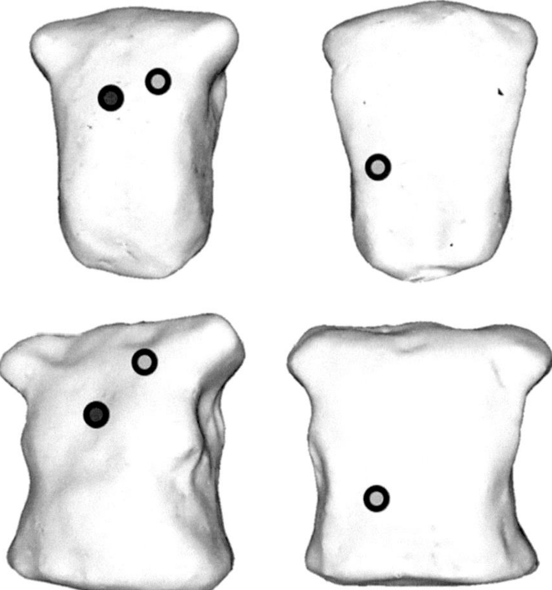

Fig. 7.11. The optimal electrode positions determined by means of computer simulation for the wearable ECG monitoring system with 3 measurement electrodes. The optimal electrode positions obtained on Patient 1 is shown in the
upper row and the one obtained on Patient 2 in the bottom row. The common electrodes among the three is marked
with a dark dot [47].

the heart, whereas for Patient 2 the third one is on the left chest but on the upper edge of the pectoralis major. In both configurations the common electrode for the two electrode pairs is the one below the fossa jugularis, which is marked with a dark dot in Fig. 7.11.

In addition, 50 myocardial infarctions, which are not included in the stochastical basis, are generated in each model for the purpose of validation. Their sizes are between 5 mm and 35 mm and their sites are randomly selected in the left ventricle including the septum. With the thresholds defined in the optimization process, i.e., $T = 0.4$ mV for Patient 1 and $T = 0.3$ mV for Patient 2, all 50 randomly simulated myocardial infarctions can be successfully detected in each model.

Discussion

As can be seen in Fig. 7.11 the optimal electrode configurations obtained on these two models are similar, but the position of the electrode on the left chest is not exactly the same. The possible reason for this is the difference of heart position between two patients (see Fig. 3.2 and Fig. 3.3). The thresholds of 0.4 mV and of 0.35 mV are at a reasonable level for the ECG measurement technique. In the validation a sensitivity of 100% is reached for both Patient 1 and Patient 2 using synthetic signals. Furthermore, the measurement noise or the inaccuracy of the measurement can be reduced by applying ST-integral.

8

Results: Inverse Problem of Electrocardiography

8.1 Inverse Problem of ECG in Realistic Environment

In the current study, ECGs with different kinds of error in the measurement and in the model as well an ECG in the so-called realistic test environment are simulated as described in Section 7.1. In the inverse problem of ECG the static model with the heart in the diastolic state and the lungs in the deflated state instead of the dynamic model is utilized. Thus, an inaccuracy is introduced into the inverse procedure due to the neglect of the heart motion and respiration. The conductivities of the myocardium, of the blood inside the heart chambers and of the lungs are at their original values in the inverse problem. It leads to a 25% inaccuracy $(1/0.8 - 1)$ comparing to the conductivity values (80% of the original values) used in the forward ECG simulation. The 64 ECG measurement electrodes remain at their original positions. Then, the transfer matrix is calculated on this model. Furthermore, 30 dB Gaussian measurement noise and 0.1 Hz baseline wander are added into the simulated ECG. Afterwards, the simulated ECGs with different kinds of error and with all errors (in realistic environment) are applied as input for the inverse problem of ECG. The source model applied in this application is epicardial potentials.

In this study the following regularization methods are tested: the Tikhonov 0-order regularization with L-curve, the Tikhonov 0-order regularization with the GCV method, the GMRes method, the TTLS regularization, the spatio-temporal Tikhonov 0-order regularization with L-curve, the spatio temporal Tikhonov 0-order regularization with the GCV method, the LSQR-Tikhonov hybrid regularization with L-curve, the LSQR-Tikhonov hybrid regularization with the GCV method, the spatio-temporal LSQR-Tikhonov hybrid regularization with L-curve and the spatio-temporal LSQR-Tikhonov hybrid regularization with the GCV method. Here, the Greensite's spatio-temporal approach is applied. The optimal value of the regularization parameter for the Tikhonov regularization and for the internal Tikhonov regularization in the hybrid methods is determined using L-curve and GCV. The iteration number for the GMRes method is set to 10. The truncation parameter for the TTLS regularization is set to 10. In the Greensite's spatio-temporal framework only the first 10 elements are considered. The iteration number for the LSQR related hybrid methods is set to 70.

The results of the inverse problem in different cases and using different regularization methods are summarized in Table 8.1 and Table 8.2 using the correlation coefficient as the benchmark of the quality of inverse solutions. In the upper panel of Table 8.1 and of Table 8.2 the correlation coefficients at R-peak and at T-peak are shown, respectively. In the bottom panel of Table 8.1

and of Table 8.2 the average correlation coefficients during QRS-complex and during T-wave are displayed, respectively. Please note that the following abbreviations are used in Table 8.1 and Table 8.2: NE (no error), NS (noise), BW (baseline wander), EC (inaccuracy in the localization of electrodes), MC (inaccuracy in the estimation of the conductivity of myocardium), BC (inaccuracy in the estimation of the conductivity of blood inside the heart chambers), LC (inaccuracy in the estimation of the conductivity of lungs), HM (heart motion), LM (lung motion) and RE (realistic environment).

Moreover, the reconstructed epicardial potentials are visualized on the heart model in the idealistic environment (see Fig. 8.1 and Fig. 8.2) and in the realistic environment (see Fig. 8.3 and Fig. 8.4). For the realistic environment the reconstructions from the second cardiac cycle are shown, where the volume of lungs reaches its maximum and the conductivity of lungs is at its minimum. The simulation serving as reference is also shown in the first row in Fig. 8.1 to Fig. 8.4. The simulated reference obtained in the dynamic model are projected onto and shown on the static heart model for the sake of clarity.

Discussion

In the inverse solutions summarized in Table 8.1 and Table 8.2 and Fig. 8.1 to Fig. 8.4 it can be observed that the results obtained with all regularization methods applied in this study are of similar quality both during QRS-complex and during T-wave in the idealistic environment. The 30 *dB* measurement noise does not lead to a considerable change in reconstruction and even improves the results slightly in some cases. It proves that the effect of measurement noise on the inverse solution can be well eliminated by regularization. Baseline wander strongly impairs the quality of inverse solutions during QRS-complex. With the presence of baseline wander the spatio-temporal version of Tikhonov regularization is not able to give a reasonable result. The 5 *mm* inaccuracy in the electrode localization results in considerable loss of quality of inverse solution, particularly for the Tikhonov regularizaiton with GCV and the spatio-temporal Tikhonov regularization with both L-curve and GCV. The 25% error in the conductivity of myocardium does not cause considerable change in the inverse solutions. However, the inaccuracy in the estimation of conductivities of blood and lungs dramatically reduces the quality of the inverse solutions obtained using the Tikhonov regularizaiton and its spatio-temporal version. The neglect of the heart motion leads to a decline of approximately 0.1 in correlation coefficient during T-wave for all regularization methods. The neglect of respiration decimates the functionality of Tikhonov regularization and its spatio-temporal version. In the realistic environment the quality of inverse solutions is reduced by about 0.15 in correlation coefficient using all regularization methods comparing to that in the idealistic environment. A detailed analysis shows that the LSQR-Tikhonov hybrid regularization with GCV is always among the top 5 in all cases and it achieves the best results both during QRS-complex and during T-wave in the realistic environment. Furthermore, all methods applying Greensite's spatio-temporal framework are much less time-consuming than the methods not applying this framework because the problem is solved only for the first 10 elements after being projected into the orthogonal temporal space instead of for all 250 (one cardiac cycle) or 1000 time instants (one respiratory cycle) in the original space.

According to the current study, it is recommended to use only those heart beats for the inverse problem that are acquired during the same respiration state of the MRI or CT dataset in order to

Table 8.1. The correlation coefficient between the inverse solution and the reference using different regularization methods in cases of different kinds of error involved as well as in the realistic environment (Part I) [46].

		NE	NS	BW	EC	MC	BC	LC	HM	LM	RE
R-peak	Tikh (LC)	0.72	0.73	0.67	0.67	0.70	0.40	0.43	0.72	0.58	0.64
	Tikh (GCV)	0.53	0.73	0.67	−0.02	0.52	0.07	0.43	0.53	−0.03	0.64
	GMRes	0.69	0.69	0.58	0.67	0.69	0.69	0.70	0.69	0.66	0.54
	TTLS	0.67	0.67	0.37	0.66	0.66	0.67	0.68	0.67	0.67	0.37
	st-Tikh (LC)	0.72	0.71	−0.04	0.62	0.69	0.50	0.47	0.68	0.26	−0.03
	st-Tikh (GCV)	0.55	0.66	0.65	−0.07	0.54	0.06	0.44	0.55	−0.03	0.63
	LSQR-Tikh (LC)	0.70	0.72	0.65	0.67	0.69	0.72	0.71	0.70	0.59	0.60
	LSQR-Tikh (GCV)	0.72	0.73	0.67	0.68	0.71	0.73	0.73	0.72	0.67	0.66
	st-LSQR-Tikh (LC)	0.73	0.74	0.41	0.69	0.72	0.73	0.75	0.72	0.68	0.52
	st-LSQR-Tikh (GCV)	0.72	0.74	0.65	0.70	0.71	0.73	0.74	0.72	0.64	0.65
QRS-complex	Tikh (LC)	0.63	0.57	0.44	0.53	0.62	0.50	0.51	0.63	0.41	0.42
	Tikh (GCV)	0.61	0.54	0.48	0.03	0.60	0.06	0.39	0.61	0.01	0.47
	GMRes	0.58	0.55	0.33	0.55	0.58	0.59	0.59	0.58	0.58	0.30
	TTLS	0.55	0.52	0.16	0.54	0.54	0.55	0.56	0.55	0.56	0.15
	st-Tikh (LC)	0.58	0.57	−0.01	0.34	0.57	0.30	0.38	0.58	0.38	0.01
	st-Tikh (GCV)	0.57	0.53	0.51	−0.02	0.56	0.06	0.36	0.57	0.03	0.47
	LSQR-Tikh (LC)	0.61	0.58	0.46	0.54	0.60	0.61	0.63	0.61	0.59	0.45
	LSQR-Tikh (GCV)	0.64	0.58	0.48	0.56	0.63	0.64	0.65	0.64	0.60	0.48
	st-LSQR-Tikh (LC)	0.60	0.57	0.25	0.49	0.60	0.59	0.60	0.61	0.58	0.30
	st-LSQR-Tikh (GCV)	0.62	0.57	0.49	0.51	0.61	0.62	0.63	0.62	0.58	0.48

Table 8.2. The correlation coefficient between the inverse solution and the reference using different regularization methods in cases of different kinds of error involved as well as in the realistic environment (Part II) [46].

		NE	NS	BW	EC	MC	BC	LC	HM	LM	RE
T-peak	Tikh (LC)	0.61	0.70	0.64	0.63	0.60	0.55	0.53	0.61	0.40	0.54
	Tikh (GCV)	0.55	0.70	0.64	−0.10	0.55	0.30	0.53	0.43	0.05	0.54
	GMRes	0.70	0.70	0.60	0.67	0.69	0.70	0.70	0.60	0.69	0.48
	TTLS	0.63	0.63	0.43	0.63	0.62	0.63	0.63	0.56	0.63	0.31
	st-Tikh (LC)	0.63	0.64	0.05	0.37	0.63	0.44	0.58	0.54	0.50	−0.05
	st-Tikh (GCV)	0.53	0.68	0.66	−0.04	0.52	0.24	0.49	0.42	0.10	0.58
	LSQR-Tikh (LC)	0.69	0.70	0.64	0.63	0.68	0.69	0.69	0.61	0.62	0.53
	LSQR-Tikh (GCV)	0.70	0.71	0.63	0.66	0.69	0.70	0.70	0.58	0.65	0.54
	st-LSQR-Tikh (LC)	0.71	0.69	0.63	0.43	0.68	0.71	0.71	0.61	0.68	0.48
	st-LSQR-Tikh (GCV)	0.69	0.69	0.69	0.54	0.69	0.69	0.69	0.58	0.69	0.60
T-wave	Tikh (LC)	0.61	0.63	0.45	0.51	0.60	0.47	0.53	0.48	0.34	0.38
	Tikh (GCV)	0.54	0.53	0.53	−0.06	0.53	0.22	0.52	0.41	0.02	0.47
	GMRes	0.65	0.64	0.44	0.62	0.63	0.65	0.65	0.57	0.65	0.37
	TTLS	0.58	0.58	0.32	0.59	0.57	0.58	0.59	0.53	0.60	0.25
	st-Tikh (LC)	0.62	0.60	0.01	0.36	0.61	0.44	0.55	0.52	0.41	−0.04
	st-Tikh (GCV)	0.56	0.63	0.60	0.02	0.55	0.19	0.52	0.43	0.06	0.52
	LSQR-Tikh (LC)	0.65	0.61	0.53	0.59	0.64	0.65	0.65	0.58	0.58	0.46
	LSQR-Tikh (GCV)	0.66	0.65	0.50	0.61	0.65	0.66	0.66	0.56	0.60	0.49
	st-LSQR-Tikh (LC)	0.66	0.66	0.47	0.43	0.65	0.65	0.66	0.59	0.64	0.37
	st-LSQR-Tikh (GCV)	0.67	0.65	0.61	0.51	0.66	0.66	0.67	0.57	0.66	0.54

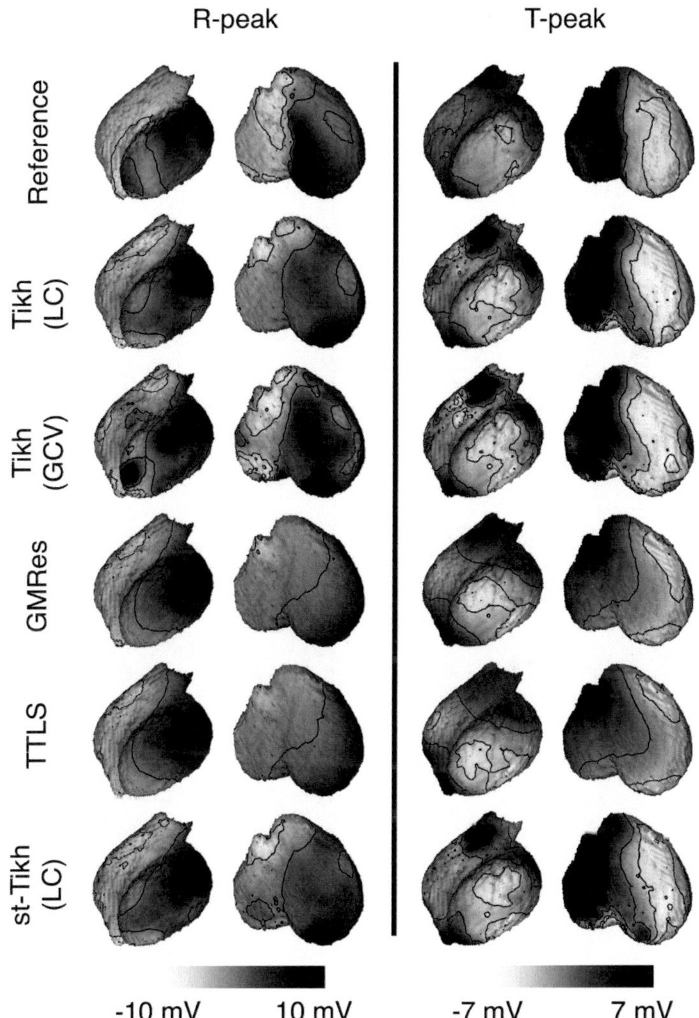

Fig. 8.1. Comparison between the reconstructed distributions of epicardial potentials obtained using different regularization methods in the case of no error involved (Part I). The results are shown at R-peak (left panel) and at T-peak (right panel). The simulation serving as reference is shown in the first row [46].

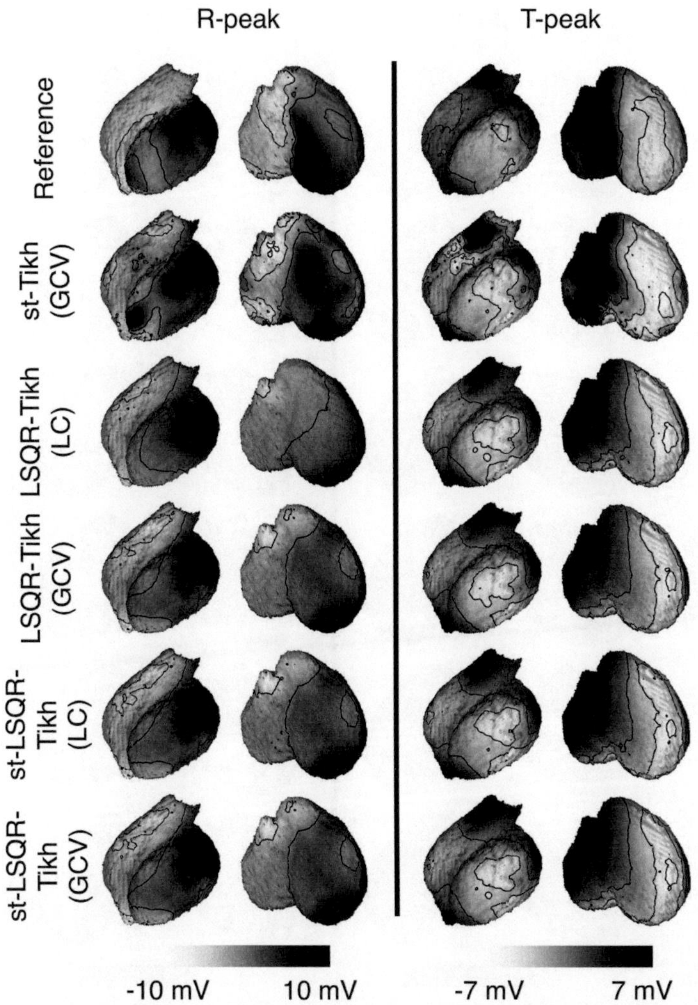

Fig. 8.2. Comparison between the reconstructed distributions of epicardial potentials obtained using different regularization methods in the case of no error involved (Part II). The results are shown at R-peak (left panel) and at T-peak (right panel). The simulation serving as reference is shown in the first row [46].

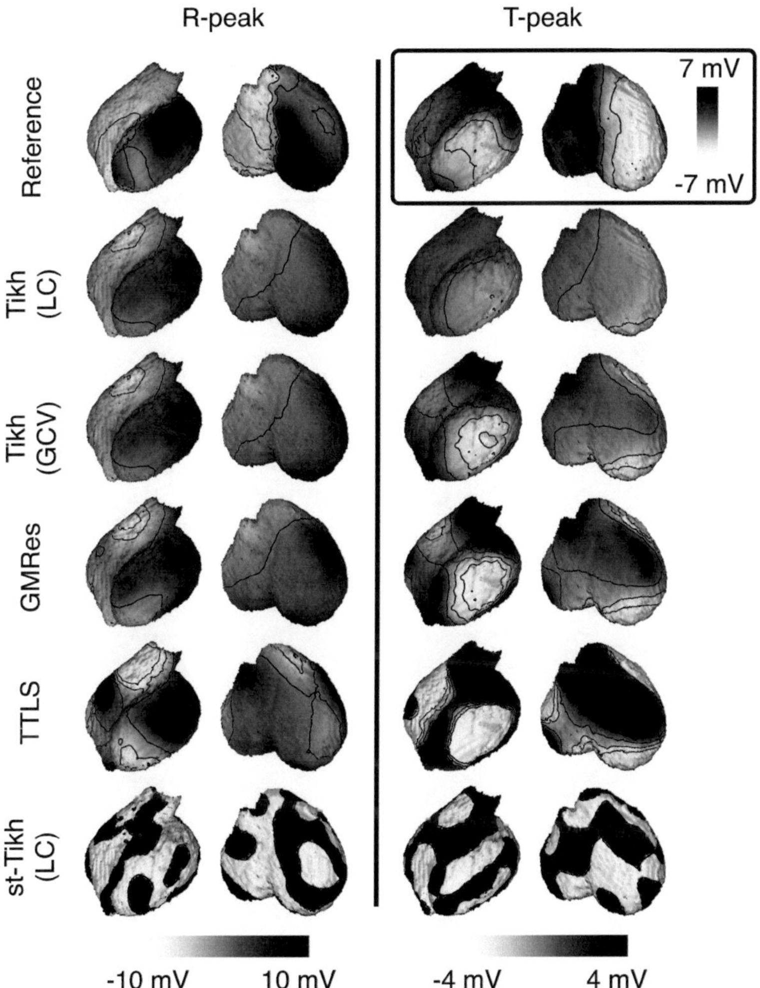

Fig. 8.3. Comparison between the reconstructed distributions of epicardial potentials obtained using different regularization methods in the realistic environment (Part I). The results are shown at R-peak (left panel) and at T-peak (right panel). The simulation serving as reference is shown in the first row [46].

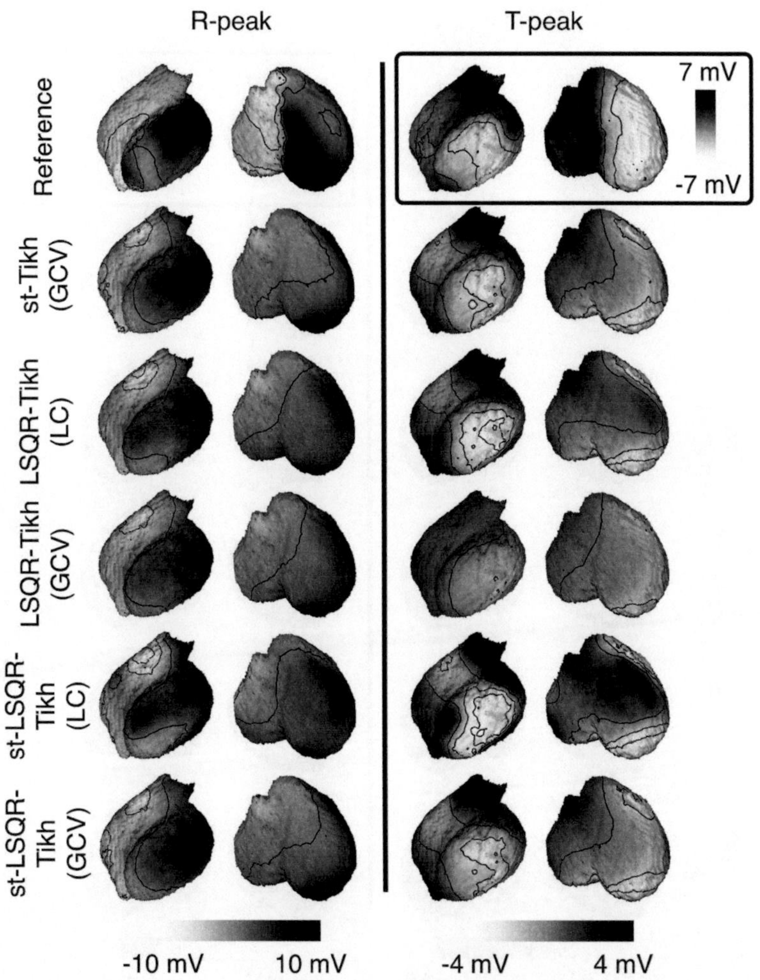

Fig. 8.4. Comparison between the reconstructed distributions of epicardial potentials obtained using different regularization methods in the realistic environment (Part II). The results are shown at R-peak (left panel) and at T-peak (right panel). The simulation serving as reference is shown in the first row [46].

get rid of the influence of respiration on the inverse solution. It is also suggested to deploy more *a priori* information in the regularization to achieve better reconstruction results, e.g., model based optimization [104, 105] and maximum *a posteriori* based regularization [106, 107, 108].

8.2 Identification of the Origin of PVC

The objective of the present investigation is to identify the origin of PVC by the solution of the inverse problem of ECG applying the integral of transmembrane voltages as source model (see Section 5.3). The Tikhonov 2nd order regularization and the MAP-based regularization are tested. In order to demonstrate the advantage of the proposed source model, the transmembrane voltages are also applied as source model to solve the inverse problem.

When the integral of transmembrane voltages is selected as source model, the integral of multi-channel ECG over the time interval between 8 *ms* before and 20 *ms* after the beginning of the PVC is calculated and applied as the input for the inverse problem of ECG. When the transmembrane voltages are applied, the inverse problem is solved frame-by-frame.

As described in Section 7.2 a stochastical basis including 243 possible PVCs in the entire left ventricle and the ventricular septum is created. The statistical property extracted from this comprehensive stochastical basis is applied as *a priori* information in the MAP-based regularization, i.e., the covariance matrix C_x of cardiac sources x is calculated from the stochastical basis (see Section 5.4.6). The covariance matrix of errors C_e is assumed to have a form of $\sigma^2 I$, where σ^2 is estimated from experience and I denotes the identity matrix. In the Tikhonov regularization the L-curve method is employed to determine the optimal value for the regularization parameter λ.

The origin of PVC is identified by the maximum in the inverse solution. If the maximum doesn't converge to one point but covers a large region, or in case more than one maximum with similar amplitude is detected, it will be considered as failure, i.e., the origin of the current PVC under consideration cannot be localized.

Synthetic ECG

In order to evaluate the feasibility of the proposed source model in localizing the origin of PVC, several simulated multichannel ECGs, which are associated with PVCs in the stochastical basis, are taken as input to the inverse problem after being contaminated by a Gaussian white measurement noise of $SNR = 30\ dB$. 4 cases are presented in Fig. 8.5 as an example. Each case is arranged in one row. Starting from the left side, the simulated data serve as ground truth shown in the first column, the inverse solutions obtained with Tikhonov regularization and MAP-based regularization using transmembrane voltages as source model are shown in the second and third columns, and the solutions obtained with Tikhonov regularization and MAP-based regularization using the integral of transmembrane voltages as source model are shown in the fourth and fifth columns. The white dot marks the maximum in the reconstruction (the origin of PVC). The localization error is shown in *mm* under the reconstruction. If the origin of one PVC could not be determined, "failed" is noted.

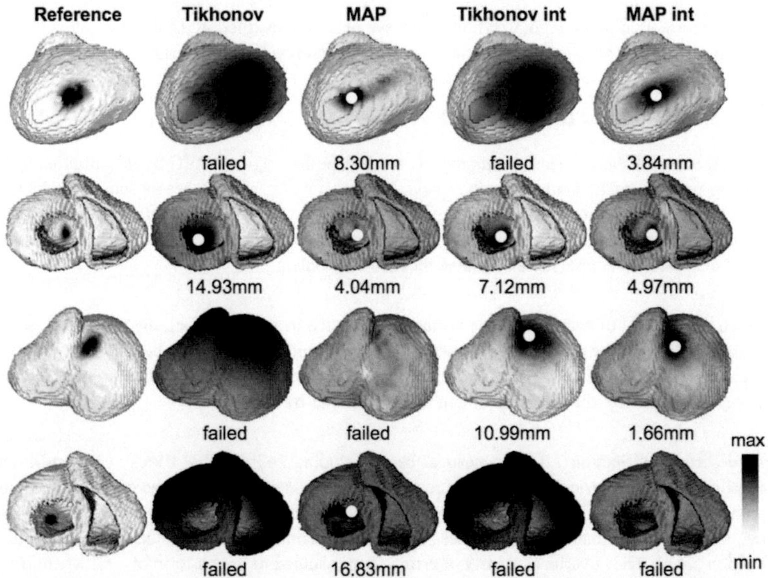

Fig. 8.5. The inverse solutions obtained from the synthetic noisy multichannel ECGs with Tikhonov regularization (the second column) and MAP-based regularization (the third column) using transmembrane voltages as source model; the inverse solutions obtained with Tikhonov regularization (the fourth column) and MAP-based regularization (the fifth column) using the integral of transmembrane voltages as source model. The white dot indicates the maximum in the reconstruction, i.e., the origin of PVC. The simulated data is shown as reference (the first column). The localization error is stated under every reconstruction. These 4 cases are taken from the stochastical basis [109].

Afterwards, 10 PVCs are randomly generated in the left ventricular wall including the septum. These PVCs do not belong to the statistical basis to build C_x. The corresponding multichannel ECGs are obtained by solving the forward problem. After adding 30 dB measurement noise these 10 multichannel ECGs are applied as input for the inverse problem of ECG. For the MAP-based regularization C_x is still extracted from the stochastical basis. 4 cases out of 10 are shown in Fig. 8.6 as an example.

Furthermore, the capability of the MAP-based regularization using the new source to differentiate the PVCs originating from endocardium, midmyocardium and epicardium is investigated. Three PVCs starting from the same segment but different depths in the ventricular wall are considered (see Fig. 7.5 the upper row). As before, the corresponding ECGs are calculated and 30 dB noise is added. The reconstructions from these three synthetic noisy ECGs using the MAP-based regularization method applying the integral of transmembrane voltages as source model are shown in Fig. 8.7 from two perspectives (endocardium and epicardium).

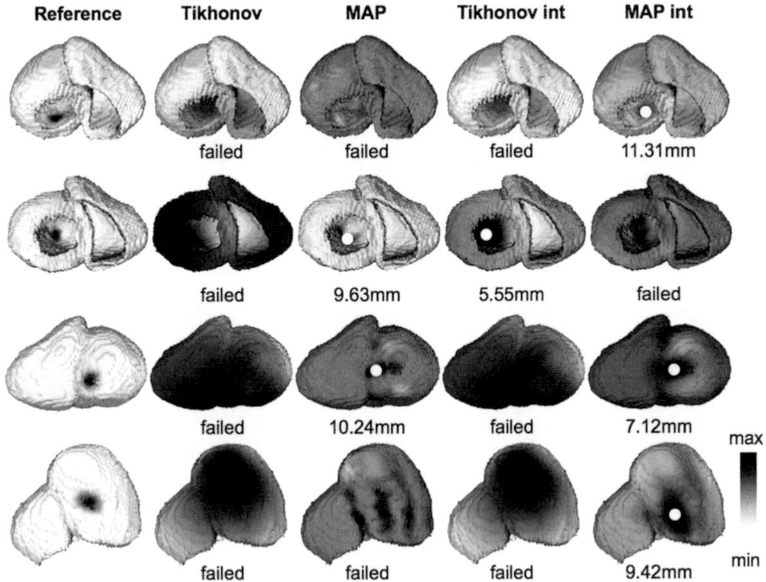

Fig. 8.6. The inverse solutions obtained from the synthetic noisy multichannel ECGs with Tikhonov regularization (the second column) and MAP-based regularization (the third column) using transmembrane voltages as source model; the inverse solutions obtained with Tikhonov regularization (the fourth column) and MAP-based regularization (the fifth column) using the integral of transmembrane voltages as source model. The white dot indicates the maximum in the reconstruction, i.e., the origin of PVC. The simulated data is shown as reference (the first column). The localization error is stated under every reconstruction. These 4 cases are not included in the stochastical basis [109].

For the estimation of the error level σ^2 is set to 1×10^{-6} for the synthetic noisy ECGs, when transmembrane voltages are chosen as source model. σ^2 is set to 5×10^{-6} for the synthetic noisy ECGs, when the integral of transmembrane voltages is selected.

Experimental ECG

Two sets of measured multichannel ECG of Patient 2 are applied to solve the inverse problem using the MAP-based regularization with the proposed source model. In the measured signals two time windows showing two different PVCs are selected. Then, the measured ECGs are synchronized with one simulated ECG: First, the measured ECG is interpolated to have the same sample rate and the same length in time as the simulated one. Second, the measured ECG is moved until its premature R-peak matches the one of the simulated ECG. The inverse problem is solved from these two experimental multichannel ECGs using the MAP-based regularization. The stochastical basis of the 243 simulated PVC is employed as *a priori* information. The variance of noise σ^2 is estimated at 1×10^{-3} for the first set of experimental data and 5×10^{-5} for the second one. The experimental multichannel ECGs after synchronization and the inverse solutions are shown

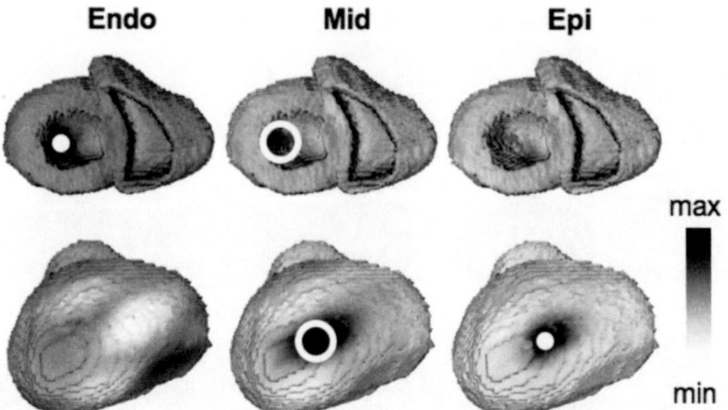

Fig. 8.7. The inverse solutions obtained from three synthetic noisy multichannel ECGs (see Fig. 7.5 the bottom row) with the MAP-based regularization using the integral of transmembrane voltages as source model. The three PVCs originate from endocardium (the left column), midmyocardium (the middle column) and epicardium (the right column). The white dot indicates the maximum in the reconstruction, i.e., the origin of PVC. The white cycle in the middle column shows the region with the largest amplitude, where the origin of PVC is included [109].

in Fig. 8.8.

Discussion

In Fig. 8.5 and Fig. 8.6 the reconstructions of 8 simulated PVCs are presented. As can be observed the Tikhonov regularization cannot deliver acceptable reconstructions. However, the application of the new source model can still help to improve the success rate and reduce the localization error. When MAP-based regularization is deployed, a significant increase in success rate is achieved: 5/8 for the conventional source model transmembrane voltages and 6/8 for the new source model the integral of transmembrane voltages. In addition, the localization accuracy is remarkably improved by applying the new source model. The MAP-regularization with the new source model achieves an average localization error of 6.4 *mm* in all 8 cases.

From the reconstructions presented in Fig. 8.7 it can be clearly seen that PVCs starting at endocardium, midmyocardium and epicardium can be distinguished using the MAP-based regularization method and applying the integral of transmembrane voltages as source model.

In the experimental study two sets of measured multichannel ECGs of the patient suffering from PVC are deployed. Due to the difference in the shape of the ECGs these two PVCs are supposed to originate from two different sites. Unique and reasonable ectopic centers are found in both cases using the MAP-based regularization method and the new source model (see Fig. 8.8). However, the lack of clinical intracardiac measurement data does not allow the further validation of the reconstructions by comparing with the real world data.

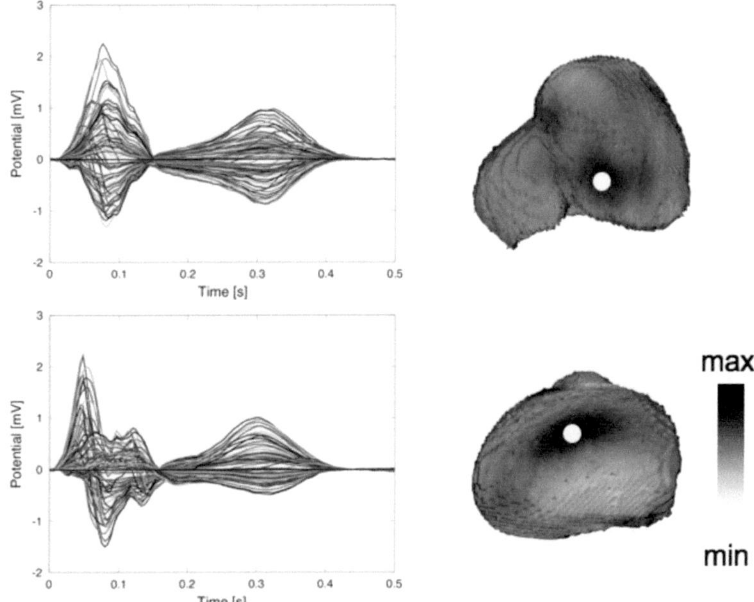

Fig. 8.8. The inverse solutions obtained from two experimental multichannel ECGs with the MAP-based regularization using the integral of transmembrane voltages as source model. The white dot indicates the maximum in the reconstruction, i.e., the origin of PVC [109].

8.3 Reconstruction of Myocardial Infarction

The current study is aimed to reconstruct myocardial infarctions by solving the inverse problem of ECG using the proposed spatio-temporal MAP-based regularization method (see Section 5.4.9). Transmembrane voltages are selected as source model for the inverse problem of ECG. Because the most significant symptom of myocardial infarction in ECG is the elevation or depression of ST-segment, the inverse problem focuses on ST-segment, i.e., only the ECG signals at the time instants within ST-segment are involved in the spatio-temporal framework. The Greensite's spatio-temporal approach is applied and the first 10 elements in the orthogonal temporal space are taken into account.

As described in Section 7.3 a stochastical basis containing 153 possible myocardial infarctions throughout the left ventricle including the septum is generated using the cellular automaton. The stochastical basis serves as *a priori* information for the spatio-temporal MAP-based regularization. The computation of the covariance matrix of C_q is done in the orthogonal temporal space (see

Eq. 5.3). The covariance matrix of errors is estimated from experience with the form of $C_\varepsilon = \sigma^2 I$.

In addition to the spatio-temporal MAP-based regularization, the inverse problem is also solved using the Tikhonov 2nd order regularization and the spatial MAP-based regularization for the sake of comparison. For the Tikhonov regularization the optimal value of the regularization parameter is determined using the L-curve method.

Synthetic ECG

In order to prove the feasibility of the spatio-temporal MAP-based regularization method in re-constructing myocardial infarctions from body surface potentials, all 153 synthetic multichannel ECGs, which are forward calculated from the simulated myocardial infarctions in the stochastical basis, are applied to the inverse problem of ECG as input data after adding a 30 dB Gaussian white measurement noise. The simulated infarctions serve as reference for evaluating the reconstructions. The correlation coefficient between the reference and the reconstruction is calculated and applied to quantify the quality of the inverse solutions. The correlation coefficients in all 153 cases are presented in Fig. 8.9 for the three regularization methods applied in the study. Out of the 153 cases, the reconstruction results of 6 cases are shown in Fig. 8.10. The transmembrane voltages in the ventricles in the middle of ST-segment are visualized on the heart surface and in a cross section of the heart for the reference and the reconstructed myocardial infarctions obtained using the three methods applied. The region with negative transmembrane voltages during ST-segment is considered the infarction region.

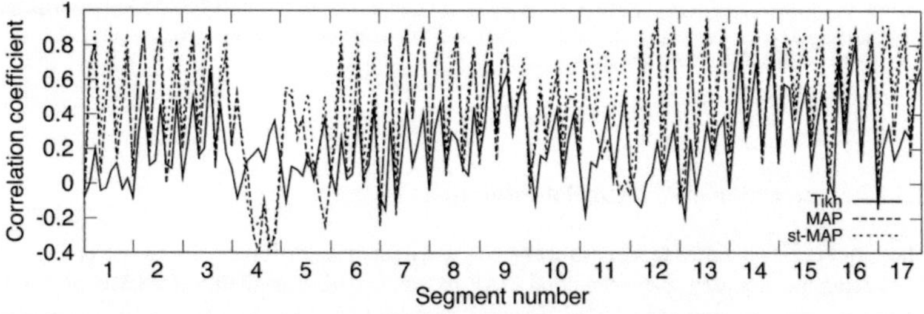

Fig. 8.9. The correlation coefficients between simulated references and the reconstructions obtained using the Tikhonov 2nd-order regularization, the spatial MAP-based regularization and the spatio-temporal MAP-based regularization. In each segment 9 myocardial infarctions are included: the first three are endocardial infarctions, the second three are transmural infarctions, and the third three are subepicardial infarctions. For each type infarctions with radii of 10 mm, 20 mm and 30 mm are simulated [103].

Subsequently, 10 myocardial infarctions are generated randomly in the left ventricular wall (including the septum) with a radius varying from 5 mm to 35 mm. These 10 myocardial infarctions are not members of the statistical basis to build C_x for the spatial MAP-based regularization and C_q for the spatio-temporal MAP-based regularization. The corresponding ECGs are calculated from these 10 myocardial infarctions and also corrupted by 30 dB Gaussian white noise. Then,

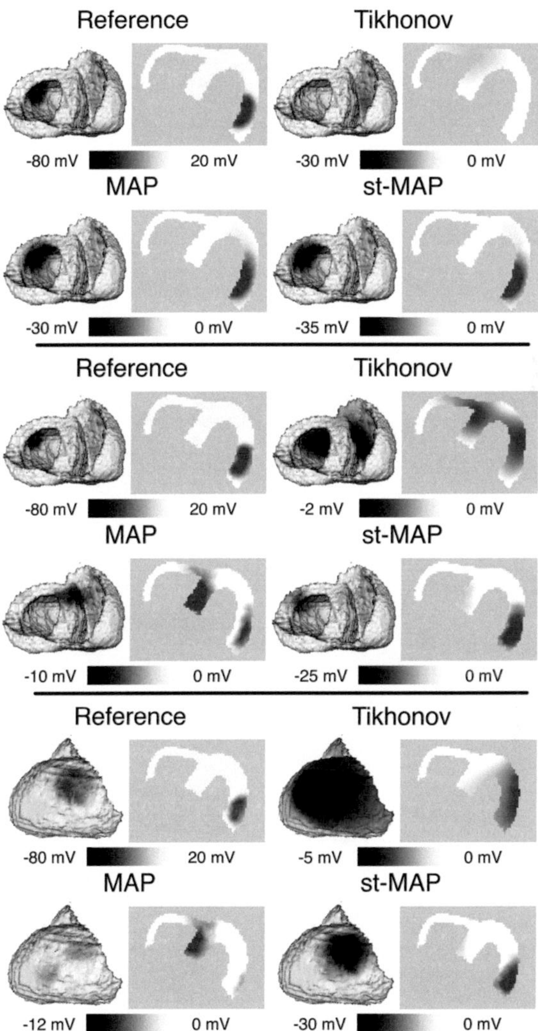

Fig. 8.10. The reconstructed myocardial infarctions obtained from 6 synthetic noisy 64-channel ECGs with the Tikhonov 2nd-order regularization, the spatial MAP-based regularization and the spatio-temporal MAP-based regularization (Part I). The simulated data are shown as reference. The reconstructions in the middle of ST-segment are shown on the heart surface and in a slice of the heart. These 6 cases are included in the stochastical basis [103].

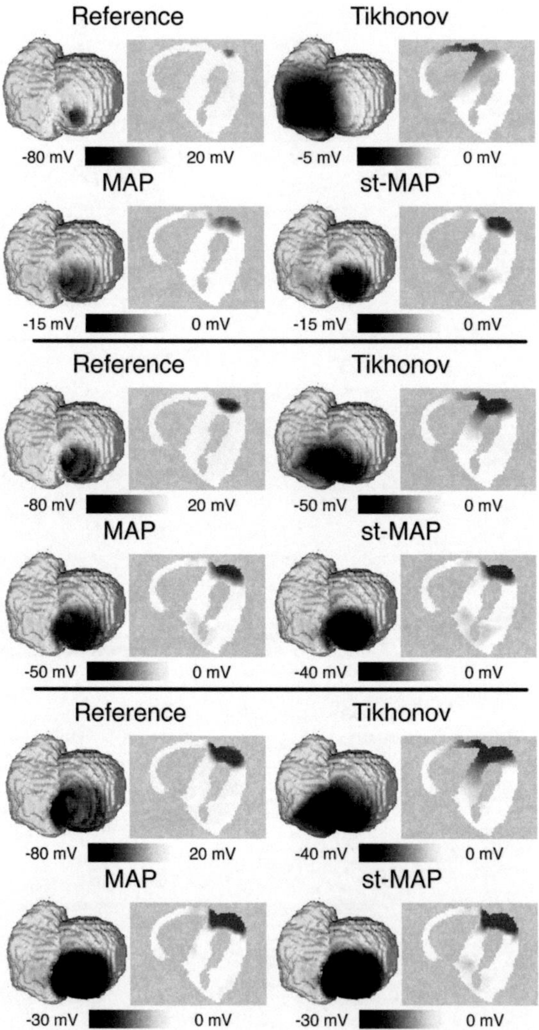

Fig. 8.11. The reconstructed myocardial infarctions obtained from 6 synthetic noisy 64-channel ECGs with the Tikhonov 2nd-order regularization, the spatial MAP-based regularization and the spatio-temporal MAP-based regularization (Part II). The simulated data are shown as reference. The reconstructions in the middle of ST-segment are shown on the heart surface and in a slice of the heart. These 6 cases are included in the stochastical basis [103].

they are applied as input for the inverse problem of ECG. For the two MAP-based regularization methods C_x and C_q are extracted from the stochastical basis as discussed above. The reconstructions obtained using the three regularization methods in three cases out of 10 are displayed in Fig. 8.12.

For the estimation of the error level in the synthetic noisy ECG σ^2 is set to 5×10^{-5} in the spatial MAP-based regularization and σ^2 is set to 1×10^{-4} in the spatio-temporal MAP-based regularization.

Experimental ECG

A set of measured multichannel ECG of Patient 1 that suffers from a myocardial infarction is deployed in the reconstruction of myocardial infarction. The noise depression and baseline wander removal are applied to the measured ECG in the prefiltering process. Then, the experimental ECG is synchronized with one simulated ECG: First, the measured ECG is scaled to have the same amplitude as the simulated one at R-peak. Second, it is interpolated to have the same length of ST-segment as the simulated one. Afterwards, the inverse problem is solved from the synchronized experimental ECG using the spatio-temporal MAP-based regularization. The *a priori* information implied in the regularization is from the same stochastical basis as the one used in the synthetic cases. In this case the variance of noise σ^2 is estimated at 0.5. The experimental 64-channel ECG after synchronization, and the reconstructed myocardial infarction are shown in Fig. 8.13. The myocardial infarction marked by a cardiologist is shown in a late enhancement MRI scan serving as the reference (see Fig. 8.13 b).

Discussion

For the inverse solutions a quantitative evaluation of reconstructions from synthetic noisy ECGs for all 153 cases included in the stochastical basis is provided in Fig. 8.9. It can be seen that the spatio-temporal MAP-based regularization achieves the best results among the three regularization methods applied. Significant improvement of the spatio-temporal MAP-based regularization over its spatial version can be observed in Segments 6, 10, 11, 15. Overall, the quality of the reconstructions of myocardial infarctions with small size, e.g., infarctions with 10 *mm* in size, is not satisfying. Myocardial infarctions in the inferior segments (Segments 4, 5, 10 and 11) are difficult to reconstruct using all three methods, since they are relatively far from the measurement electrodes. Because myocardial infarction leads to the change of local tissue property and this change isn't considered in the computation of the transfer matrix (the location of infarction is unknown), a certain amount of modeling error is introduced in the system. For this reason, the Tikhonov regularization is not able to deliver good results as shown in Fig. 8.10. The spatial MAP-based regularization fails to reconstruct the transmural and subepicardial infarctions in the mid inferior segment (Segment 11). The spatio-temporal MAP-based regularization provides accurate reconstructions (both location and size) in all 6 cases that are included in the stochastical basis. For the randomly generated myocardial infarctions that are not included in the stochastical basis the same conclusions are reached (see Fig. 8.12).

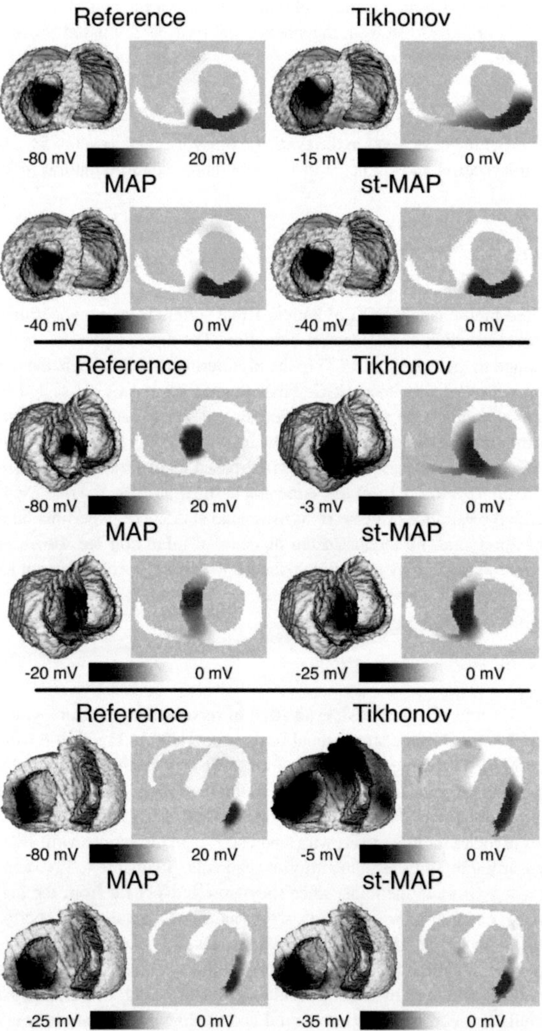

Fig. 8.12. The reconstructed myocardial infarctions obtained from 3 synthetic noisy 64-channel ECGs with the Tikhonov 2nd-order regularization, the spatial MAP-based regularization and the spatio-temporal MAP-based regularization. The simulated data are shown as reference. The reconstructions in the middle of ST-segment are shown on the heart surface and in a slice of the heart. These 3 cases are randomly generated and not included in the stochastical basis [103].

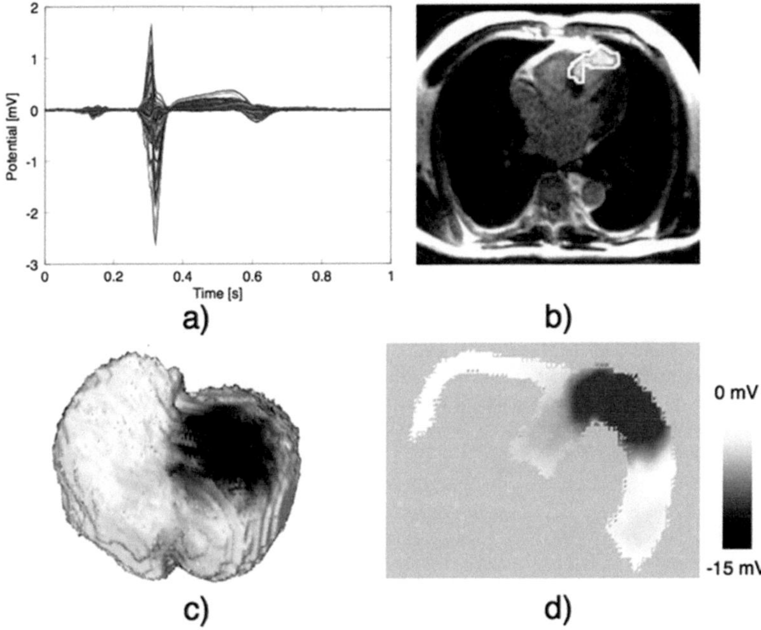

Fig. 8.13. The experimental 64-channel ECG (a), the myocardial infarction marked in the late enhancement MRI by the cardiologist (b) and the reconstruction from the experimental ECG using the spatio-temporal MAP-based regularization in the middle of ST-segment shown on the heart surface (c) and in a slice of the heart (d) [103].

In the experimental case a significant ST-elevation is shown in the 64-channel ECG signal from the patient with myocardial infarction (see Fig. 8.13 a). A unique negative region is found in the reconstructed transmembrane voltages using the spatio-temporal MAP-based regularization. The reconstruction shows a myocardial infarction in the anterior side of the heart, and the septum is also partially affected (see Fig. 8.13 c and d). The reconstruction result shows good accordance with the diagnosis of an experienced cardiologist shown on the late enhancement MRI (see Fig. 8.13 b).

8.4 Optimization of Electrode Positions for a BSPM System

The aim of the current investigation is to optimize a 64-channel BSPM system used for the inverse problem of ECG. This optimization is concentrated on one specific application: the reconstruction of myocardial infarctions. As described in Section 5.5 the optimization strategy is based on the analysis of the left (spatial) singular vector of the stochastical basis including the different possibilities of myocardial infarctions in the left ventricle (see Section 7.3). 663 nodes distributed evenly on the torso surface are selected for the analysis. The first 5 left singular vectors are taken into account. In order to provide a clear overview of the analysis results, the singular vectors are normalized after taking absolute value (see Fig. 8.14 a to e) and the sum of them is calculated (see Fig. 8.14 f).

The original electrode configuration is shown in Fig. 8.15 a. It consists of 7 strips of electrodes. 4 strips (48 electrodes) are located bilateral symmetrically on the front side of thorax and 3 other strips (16 electrodes) are placed on the left lateral side of thorax. In the optimized configuration (see Fig. 8.15 b) 52 electrodes are located on the front side of thorax, 4 electrodes on the back and 8 electrodes on the left shoulder. The electrodes cover the regions where the strongest signals appearing in the sum of the first 5 left singular vectors (see Fig. 8.14 f).

For the original and the optimized electrode configurations the transfer matrices between the cardiac sources (transmembrane voltages) and the measurements (ECG signals at the measurement electrodes) are calculated. In addition, the condition number and singular values of these two transfer matrices are calculated to evaluate the electrode configurations with respect to the nullspace theory. The condition number of the transfer matrix of the original system is 6.25^{+06} and the optimized system reaches a condition number of 2.26^{+06}. The singular values of two configurations are plotted in Fig. 8.16.

Afterwards, the inverse problem of ECG is solved using the original and optimized electrode configurations, separately. The ECG signals associated with the simulated 153 myocardial infarctions in the stochastical basis are obtained by solving the forward problem of ECG. In this study the integral of transmembrane voltages is chosen as source model as described in Section 5.3. Because the reconstruction of myocardial infarctions focuses on ST-segment, the integral of the simulated ECG over ST-segment is calculated and taken as the input for the inverse problem. Accordingly, the output of inverse problem is the integral of transmembrane voltages over ST-segment. Two regularization methods are utilized in the current study. The first on is the Tikhonov 2nd-order regularization with the L-curve method. The second one is the MAP-based regularization incorporating the simulation results in the stochastical basis as *a priori* information.

The correlation coefficients between the simulated references and the reconstructions obtained with Tikhonov 2nd-order regularization using the original and optimized electrode configurations for all myocardial infarctions in the stochastical basis are plotted in Fig. 8.17. As an example, the reconstructions of the basal inferoseptal transmural infarction (Segment 3) with a radius of 30 *mm* and of the mid anterolateral transmural infarction (Segment 12) with a radius of 20 *mm* are shown in Fig. 8.18.

Similarly, the correlation coefficients between the simulated references and the reconstructions obtained with the MAP-based regularization using the original and optimized electrode configurations for all 153 myocardial infarctions are plotted in Fig. 8.19. As an example, the reconstructions of the basal inferior transmural infarction (Segment 4) with a radius of 30 *mm* and of the mid inferoseptal infarction (Segment 9) with a radius of 30 *mm* are shown in Fig. 8.20.

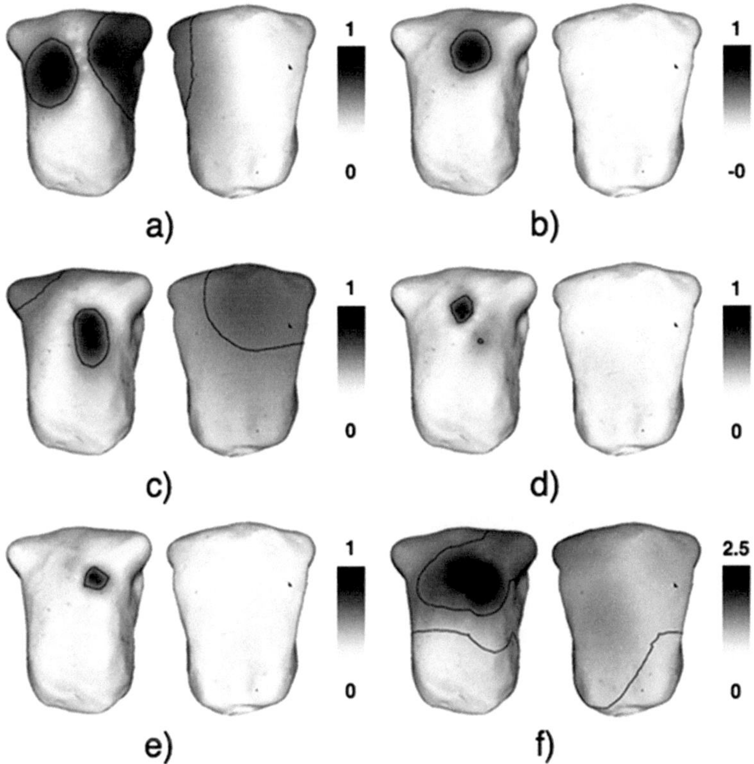

Fig. 8.14. The first 5 left singular vectors of the 153 simulated ST-integral maps after taking absolute value and normalization (a) (b) (c) (d) (e); the sum of the first 5 normalized left singular vectors after taking absolute value (f) [110].

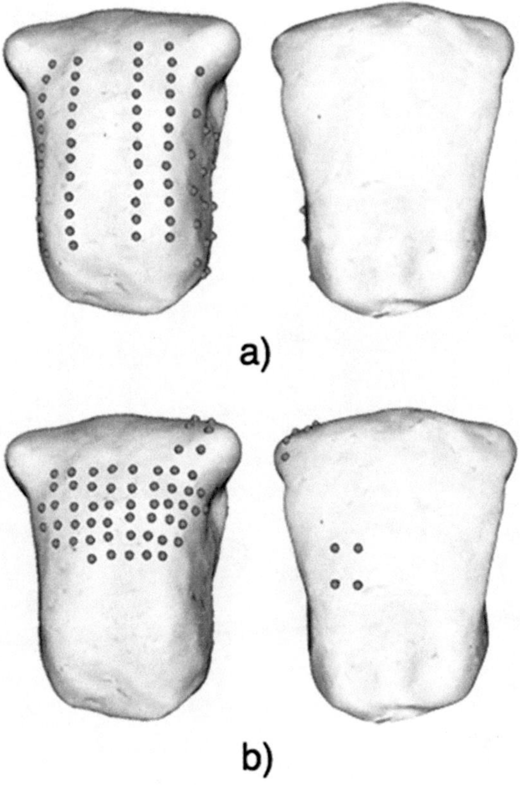

Fig. 8.15. The original electrode configuration (a) and the optimized electrode configuration of the 64-channel ECG system [110].

Discussion

The original electrode configuration does not provide a satisfactory coverage of the first 5 important left singular vectors as shown in Fig. 8.14. The electrodes are not sufficiently dense in the precordial region and no electrodes are placed on the left shoulder and on the back. It also can be seen that those measurement electrodes located on the lower part of the thorax are redundant. Furthermore, the condition number of the original BSPM system is about 3 times higher than that of the optimized BSPM system. It indicates that the original system is more ill-posed than the optimized system. The comparison of the steepness of the slope of singular values as a function of index (within the first 40 singular values) in Fig. 8.16 also shows the same fact. In addition, this conclusion is true not only for the reconstruction of myocardial infarctions, but also for the inverse problem of ECG in general.

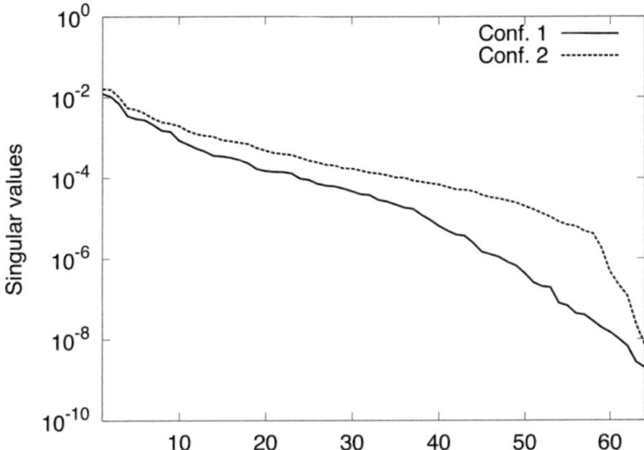

Fig. 8.16. Singular values of the transfer matrices for the original electrode configuration (Conf. 1) and the optimized electrode configuration (Conf. 2) [110].

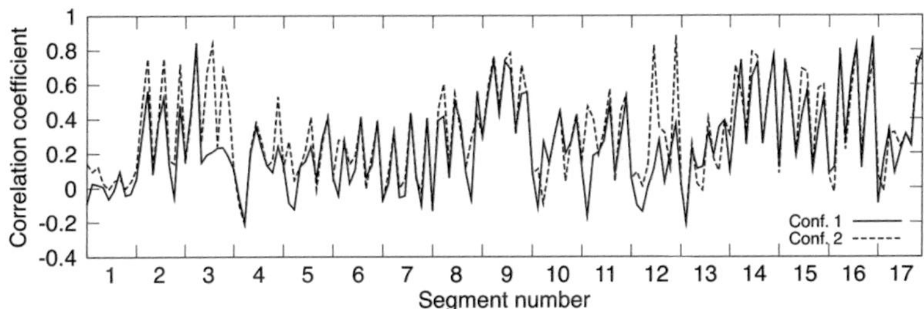

Fig. 8.17. The correlation between simulated references and the reconstructions with the Tikhonov 2nd-order regularization using the original electrode configuration (Conf. 1) and the optimized electrode configuration (Conf. 2). In each segment 9 myocardial infarctions are included: the first three are endocardial infarctions, the second three are transmural infarctions, and the third three are subepicardial infarctions. For each type infarctions with radii of 10 *mm*, 20 *mm* and 30 *mm* are simulated [110].

For the reason discussed in Section 8.3, Tikhonov regularization cannot offer satisfying reconstructions of myocardial infarctions in general. Nevertheless, the optimized electrode configuration shows significant improvement in the reconstruction quality against the original electrode configuration in almost all segments (see Fig. 8.17), particularly in Segment 3 (basal inferoseptal), Segment 11 (mid inferolateral) and Segment 12 (mid anterolateral) (see Fig. 8.17 and Fig. 8.18).

By comparing Fig. 8.19 and Fig. 8.9 it can be observed that the reconstructions obtained using the new source model the "integral of transmembrane voltages" outperformed those obtained using the conventional source model "transmembrane voltages", when the MAP-based regularization is

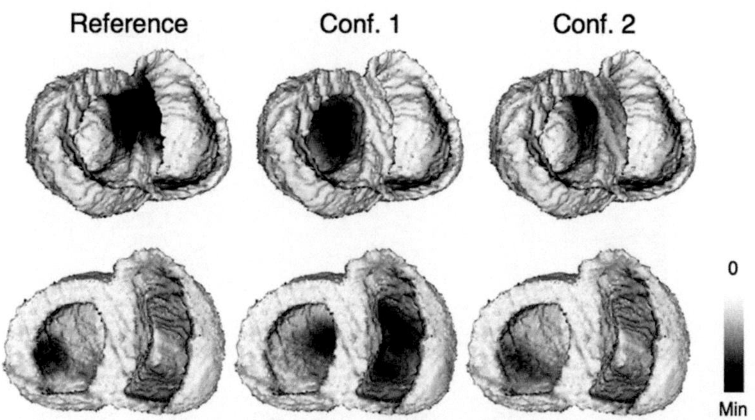

Fig. 8.18. The reconstructed myocardial infarctions with the Tikhonov 2nd-order regularization using the original electrode configuration (Conf. 1) and the optimized electrode configuration (Conf. 2). The simulations as reference are shown in the first column. The basal inferoseptal transmural infarction (Segment 3) with a radius of 30 *mm* (top) and the mid anterolateral transmural infarction (Segment 12) with a radius of 20 *mm* (bottom) are shown [110].

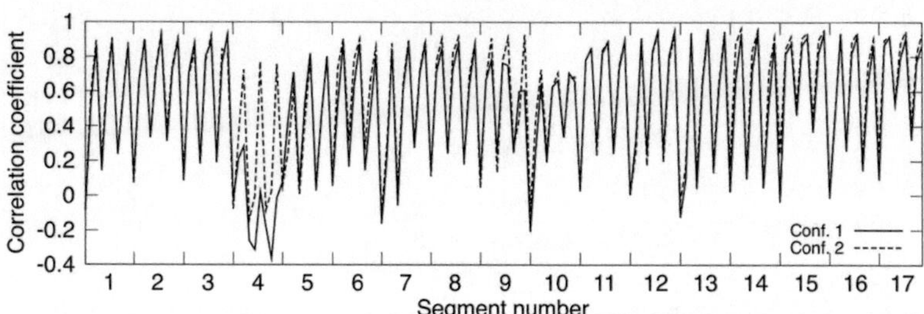

Fig. 8.19. The correlation between simulated references and the reconstructions with the MAP-based regularization using the original electrode configuration (Conf. 1) and the optimized electrode configuration (Conf. 2). In each segment 9 myocardial infarctions are included: the first three are endocardial infarctions, the second three are transmural infarctions, and the third three are subepicardial infarctions. For each type infarctions with radii of 10 *mm*, 20 *mm* and 30 *mm* are simulated [110].

applied. Furthermore, the optimized electrode configuration effectively improves the quality of the reconstructions in the inferior segments, especially in Segment 4 (basal inferior) and Segment 9 (mid inferoseptal) (see Fig. 8.19 and Fig. 8.20).

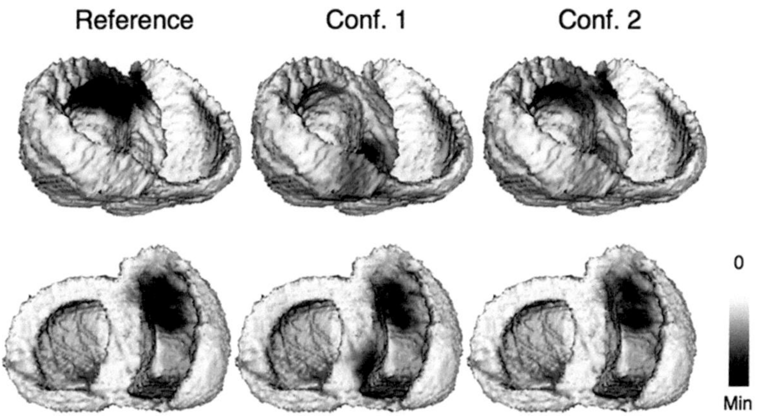

Fig. 8.20. The reconstructed myocardial infarctions with the MAP-based regularization using the original electrode configuration (Conf. 1) and the optimized electrode configuration (Conf. 2). The simulations as reference are shown in the first column. The basal inferior transmural infarction (Segment 4) with a radius of 30 *mm* (top) and the mid inferoseptal infarction (Segment 9) with a radius of 30 *mm* (bottom) are shown [110].

9

Results: Impedance Based Catheter Positioning System

As described in Section 6.2 the proposed catheter positioning system is based on the quasi-linearity between the normalized currents measured at the electrode-patches on the body surface and the position of electrodes in the heart chamber. The current simulation study is performed on the male Visible Human model (see Fig. 3.1).

9.1 Linearity Test

The following two tests are aimed to demonstrate the quasi-linearity between the normalized currents and the electrode positions.

The first test is performed in the right ventricle. An electrode moves along the y direction as shown in Fig. 9.1 a. The total length of the movement is 28 *mm*. At 15 electrode positions with a step of 2 *mm* the currents at the patches are recorded. The relationship between the currents and the electrode positions is shown in Fig. 9.1 b. After applying the normalization on the currents a linear relationship can be observed in Fig. 9.1 c. In order to display the linearity of the relationship more clearly, the least-square fitting estimated from all the measurement data points is plotted (see Fig. 9.1 d, e and f).

The second test is made to find out the change of the linearity when the electrode is touching the endocardium slightly, being docked onto the endocardium and even being embedded into the myocardium. An electrode moves along the x direction towards the septum in the right ventricle (see Fig. 9.2 a). The electrode is moved by 29 *mm* with a step of 1 *mm* from the initial position until it is deeply embedded into the septum. Although the currents decrease dramatically when the electrode is touching the endocardial surface (see Fig. 9.2 b), no significant change is shown in the normalized currents (see Fig. 9.2 c). When the electrode is entirely embedded in the myocardium, which would not happen in the clinical intracardiac mapping normally, the linearity of the normalized currents cannot persist (see Fig. 9.2 c to f) [100].

9.2 Electrode Localization

After proving the linear relationship between the normalized currents and the electrode positions several simulation studies are conducted to suggest appropriate calibration configuration and determine the localization error in different measurement scenarios as well as investigate the influ-

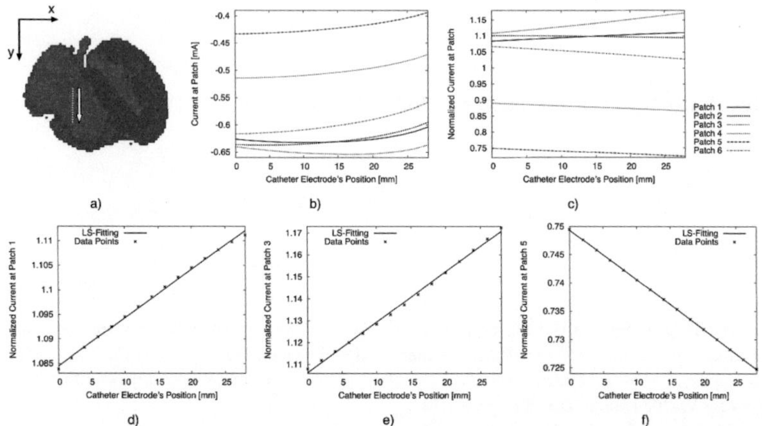

Fig. 9.1. Movement of an electrode along the y direction in the right ventricle (a). Change of the currents (b) as well as of the normalized currents (c) measured at 6 patches with the movement of the electrode. The degree of the linearity of the relationship between the normalized current at patch 1 (d), patch 3 (e) and patch 5 (f) and the electrode position is also shown, respectively [100].

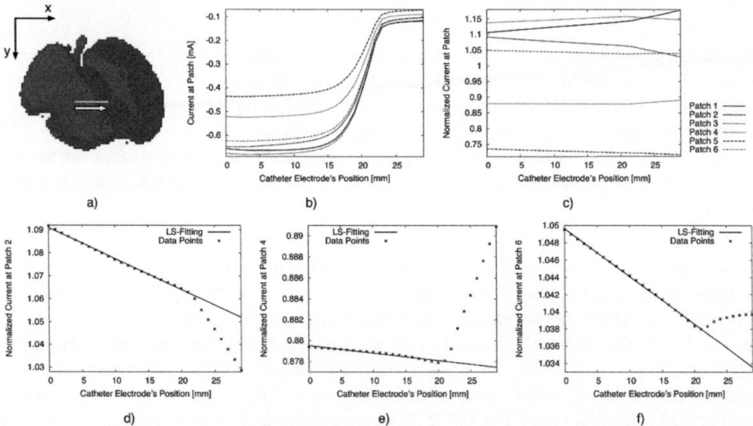

Fig. 9.2. Movement of an electrode towards the septum along the x direction (a). Change of the currents (b) as well as of the normalized currents (c) measured at 6 patches with the movement of the electrode. The degree of the linearity of the relationship between the normalized current at patch 2 (d), patch 4 (e) and patch 6 (f) and the electrode position is also shown, respectively. In (d) (e) and (f) the least-square fitting is calculated from the first 18 data points. The electrode reaches the septum at $x = 18$ mm [100].

ence of the inherent localization inaccuracy of the localisable catheter on the localization results.

Scenario 1

In the first study the electrode localization is performed in a cube, which represents a small environment in the heart chamber. 3 different sizes of the cube (12 *mm*, 16 *mm* and 20 *mm* side length) are applied in the left atrium, the right atrium and the right ventricle (see Fig. 9.3 a, b and c). Each cube includes 35 electrode positions. Among them 8 electrode positions are known and the other 27 electrode positions are to be determined. In the first calibration configuration the 8 vertex positions are known (see Fig. 9.3 d) and in the second configuration 8 electrode positions inside the cube are given (see Fig. 9.3 e). Due to the limited blood volume in the left atrium and ventricle the investigation with the cube with a side length of 20 *mm* is not performed in the left atrium, and the left ventricle is excluded from the entire study.

The results of the electrode localization are summarized in Table 9.1. The first column shows the side length of the cube and the first row gives the chamber where the study is made. The local-

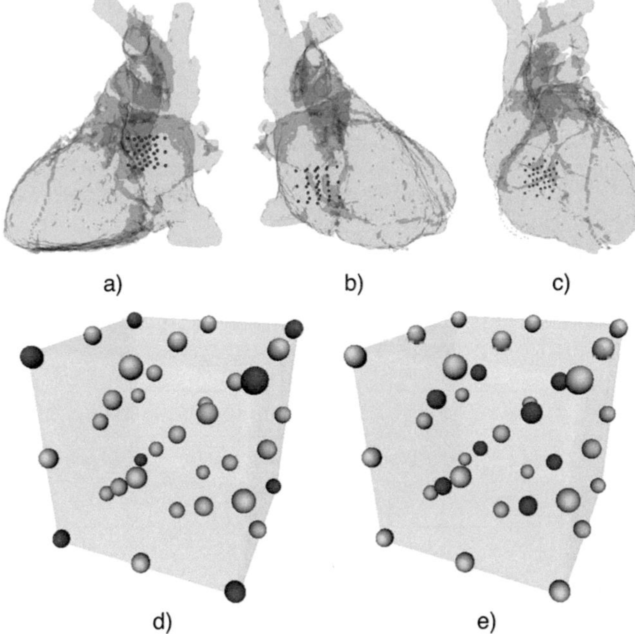

Fig. 9.3. The cube containing 35 electrodes shown in the left atrium (a), the right atrium (b) and the right ventricle (c). 8 electrode positions (dark-colored) are given by the localisable catheter and the other 27 electrodes (light-colored) are to be localized (d and e) [100].

ization error is listed in the form of mean $\pm$ standard deviation. For each cube size the first line of the localization error is associated with the first calibration configuration and the second row is for the second configuration. It can be seen that higher accuracy is achieved by using the first calibration configuration. Therefore, it suggests that the system should be calibrated in a certain region and the localization is then done within the calibrated region. The first calibration configuration is then used in the next investigation in this scenario.

Table 9.1. Localization error in different chambers of the heart (left column: side length of the calibrated cubic region) [100]

	LA	RA	RV
12 mm	0.458 ± 0.230 mm	0.395 ± 0.246 mm	0.309 ± 0.193 mm
	0.534 ± 0.486 mm	0.440 ± 0.376 mm	0.362 ± 0.240 mm
16 mm	1.053 ± 0.455 mm	0.616 ± 0.402 mm	0.531 ± 0.324 mm
	1.016 ± 0.840 mm	0.740 ± 0.371 mm	0.630 ± 0.386 mm
20 mm	—	0.936 ± 0.630 mm	0.839 ± 0.506 mm
	—	1.151 ± 0.453 mm	0.993 ± 0.593 mm

Because the currently existing localisable catheter only has a limited localization inaccuracy, a random initial localization error of up to 2 *mm* is added to the location information, which is deployed in the system calibration and accordingly in the construction of the transfer matrix A. The results of electrode localization involving the inherent localization inaccuracy are shown in Table 9.2 [100].

Table 9.2. Localization error in different chambers of the heart with the inherent inaccuracy of the localisable catheter used for the calibration included (left column: side length of the calibrated cubic region) [100]

	LA	RA	RV
12 mm	0.772 ± 0.336 mm	0.733 ± 0.313 mm	0.693 ± 0.296 mm
16 mm	1.205 ± 0.533 mm	0.883 ± 0.431 mm	0.818 ± 0.387 mm
20 mm	—	1.139 ± 0.630 mm	1.048 ± 0.540 mm

Scenario 2

Another important scenario to be investigated is that all the electrodes are docked onto or placed very close to the cardiac tissue. It corresponds to the situation that often occurs during catheter ablation.

In this simulation study a set of electrodes are located on the right side of the ventricular septum. Most of the electrodes are docked onto the septum and some of them are very close (in a distance smaller than 1 *mm*) to the septum as shown in Fig. 9.4 a. Two cases are under consideration. The first one consists of 36 electrodes in a 10 *mm* × 10 *mm* area (see Fig. 9.4 b and c) and the second one has 64 electrodes in a 14 *mm* × 14 *mm* area (see Fig. 9.4 d and e). For each case two calibration configurations are used: the first one has 4 known electrode positions (see Fig. 9.4 b and d) and the second one has 8 known electrode positions (see Fig. 9.4 c and e). The localization results without and with the influence of the inherent localization inaccuracy (random error up to 2 *mm*) are summarized in Table 9.3, where the first column is the side length of the calibrated area and the first row indicates whether the inherent localization inaccuracy is included. For each side length the results presented in the first row is associated with the first calibration configuration

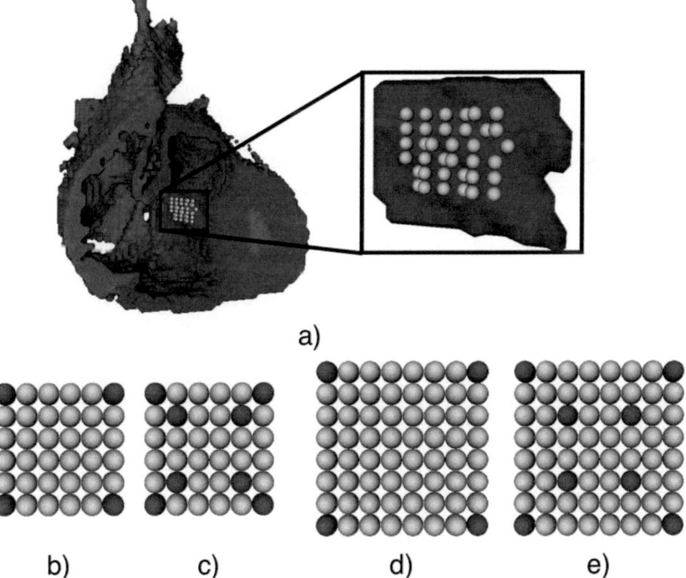

Fig. 9.4. A set of electrodes placed on the right side of the ventricular septum (a). 36 electrodes in a 10 *mm* × 10 *mm* area (b and c) and 64 electrodes in a 14 *mm* × 14 *mm* area (d and e). The positions of the dark-colored electrodes are given by the localisable catheter and the light-colored electrodes are to be localized [100].

and the second row corresponds to the second calibration configuration [100].

Table 9.3. Localization error of electrodes placed on endocardial surface without and with inherent localization inaccuracy of localisable catheter used for calibration (left column: side length of calibrated area) [100]

	w/o inh. loc. inaccuracy	with inh. loc. inaccuracy
10 *mm*	0.546 ± 0.299 *mm*	0.971 ± 0.379 *mm*
	0.287 ± 0.134*mm*	0.951 ± 0.404*mm*
14 *mm*	0.723 ± 0.357 *mm*	1.101 ± 0.407 *mm*
	0.398 ± 0.206 *mm*	0.988 ± 0.483 *mm*

Discussion

The linearity test in the human heart clearly shows that the linear relationship between the normalized currents and the electrode positions as long as the catheter electrode is "visible" in a heart chamber (not completely embedded in the cardiac tissue) as shown in Fig. 9.1 and Fig. 9.2. This ensures the usability of the proposed method of catheter localization.

In the first scenario, where all electrodes are floating in the blood, the localization error can be kept at around 1 *mm* in a region with a side length of 20 *mm* with the inherent localization inaccuracy of the existing system (see Table 9.2). In the second scenario, where all electrodes are docked onto or very close to the cardiac tissue, the localization error is around 1 *mm* in a area of up to 14 *mm* × 14 *mm* under the influence of the inherent localization inaccuracy (see Table 9.3). Overall, the localization accuracy of the proposed impedance based system is very good.

As aforementioned, the size and shape of the catheter electrode, the amount of the voltage applied at the catheter electrode and the size of the electrode-patches on the body surface are no more critical after the employment of the normalized currents. Therefore, the proposed impedance based catheter positioning system is very flexible and can cooperate with catheters manufactured by different companies. The voltage between electrode and patches can be kept as low as possible to generate harmless small currents that are allowed in the human body. Furthermore, AC voltages with different frequencies can be used for different catheter electrodes to enable the localization of all electrodes at the same time, i.e., frequency-multiplexing.

Outlook

Cardiac Modeling and Forward Simulation

In the future work the cellular automaton and the programs used for the forward ECG simulation will be further improved.

- The apico-basal dispersion of the cell properties will be included in the cellular automaton.
- New cardiac cell models will be built into the cellular automaton.
- A new option will be added to the cellular automaton to facilitate the selection of different cardiac cell models used for the simulation.
- Simulation of myocardial ischemia and infarction in different stages will be implemented in the cellular automaton or directly using the cardiac cell model.
- The forward ECG simulation with myocardial ischemia and infarction will be implemented in dynamic model.
- Realistic dynamic heart model based on 4D MRI data will be developed.
- The performance of the cellular automaton and of the forward simulation will be improved by applying parallel computing techniques and the specialized toolkit for scientific computation.

Inverse Problem of Electrocardiography

The following suggestions are made for the further investigations in the project of the inverse problem of electrocardiography.

- More *a priori* information should be implied in the solution of the inverse problem of ECG, because the methods that only use general information in the regularization cannot provide reconstructions with a sufficient quality.
- The inverse problem of ECG should rather be applied to a disease than be investigated in a general case like the sinus rhythm.
- The validation of inverse solutions will be enabled through the collaboration with clinical partners, who can provide experimental measurement data and professional medical advices.
- Parameter choice methods like "L-curve" or GCV can be implemented or introduced for TTLS, LSQR and MAP-based regularization techniques to allow the determination of optimal value for regularization or truncation parameters.
- Regularization methods should be tested and evaluated in the realistic environment including all kinds of error rather than in a test environment with only measurement noise included.

Impedance Based Catheter Positioning System

The current simulation study is done in a static model, which neglects the influence of the heart motion and the respiration on the localization of intracardiac electrodes. In the further investigation the dynamic model will be deployed in the simulation study. Through the simulation study adequate method to reduce the localization inaccuracy introduced by the heart and lung dynamics will be developed.

References

1. "Diagram of the human heart." Wikipedia.org.

2. A. Despopoulos and S. Silbernagl, *Color Atlas of Physiology*. Stuttgart; New York: Georg Thieme Verlag, 5 ed., 2001.

3. R. P. Jaakko Malmivuo, *Bioelectromagnetism*. Oxford University Press, Inc., 1995.

4. G. R. Li, J. Feng, L. Yue, and M. Carrier, "Transmural heterogeneity of action potentials and I_{to1} in myocytes isolated from the human right ventricle," *American Journal of Physiology*, vol. 275, pp. 369–377, 1998.

5. A. Bolz and W. Urbaszek, *Technik in der Kardiologie*. Berlin: Springer, 2002.

6. O. Skipa, *Linear Inverse Problem of Electrocardiography: Epicardial Potentials and Transmembrane Voltages*. PhD thesis, Institute of Biomedical Engineering, Universität Karlsruhe (TH), 2004.

7. "Visible human project." National Library of Medicine, Bethesda, USA.

8. F. B. Sachse, *Computational Cardiology. Modeling of Anatomy, Electrophysiology, and Mechanics*, vol. 2966 of *Lecture Notes in Computer Science*. Springer Verlag, Berlin, 2004.

9. A. Bowyer, "Computing dirichlet tesselations," *The Computer Journal*, vol. 24, no. 2, pp. 162–166, 1981.

10. D. F. Watson, "Computing the n-dimensional delaunay tesselation with application to voronoi polytopes," *The Computer Journal*, vol. 24, no. 2, pp. 167–172, 1981.

11. M. Müller, *Berechnung dreidimensionaler Strömungsfelder der transthorkalen Defibrillation im menschlichen Körper mit der Finite-Elemente-Methode*. PhD thesis, Institute of Biomedical Engineering, Universität Karlsruhe (TH), 1998.

12. O. C. Zienkiewicz and J. Z. Zhu, "The superconvergent patch recovery and a posteriori error estimates. part 1: The recovery technique," *International Journal for Numerical Methods in Engineering*, vol. 33, pp. 1331–1364, 1992.

13. O. C. Zienkiewicz and J. Z. Zhu, "The superconvergent patch recovery and a posteriori error estimates. part 2: Error estimates and adaptivity," *International Journal for Numerical Methods in Engineering*, vol. 33, pp. 1365–1382, 1992.

14. C. Gabriel, S. Gabriel, and E. Corthout, "The dielectric properties of biological tissues: I. Literature survey," *Physics in Medicine and Biology*, vol. 41, no. 11, pp. 2231–2250, 1996.

15. S. Gabriel, R. W. Lau, and C. Gabriel, "The dielectric properties of biological tissues: II. Measurements in the frequency range 10 Hz to 20 GHz," *Physics in Medicine and Biology*, vol. 41, no. 11, pp. 2251–2269, 1996.

16. S. Gabriel, R. W. Lau, and C. Gabriel, "The dielectric properties of biological tissues: III. Parametric models for the dielectric spectrum of tissues," *Physics in Medicine and Biology*, vol. 41, no. 11, pp. 2271–2293, 1996.

17. D. Farina and O. Dössel, "Influence of cardiac activity in midmyocardial cells on resulting ECG: Simulation study," in *Proceedings of Biomedizinische Technik*, 2006.

18. Y. Jiang, D. Farina, and O. Dössel, "Effect of heart motion on the solutions of forward and inverse electrocardiographic problem - a simulation study," in *Proceedings of Computers in Cardiology*, vol. 35, pp. 365–368, 2008.

19. Y. Jiang, Y. Meng, D. Farina, and O. Dössel, "Effect of respiration on the solutions of forward and inverse electrocardiographic problems - a simulation study," in *Proceedings of Computers in Cardiology (Rosanna Degani Young Investigator Award Nomination)*, vol. 36, pp. 17–20, 2009.

20. D. Farina, *Forward and Inverse Problems of Electrocardiography : Clinical Investigations*. PhD thesis, Institute of Biomedical Engineering, Universität Karlsruhe (TH), 2008.

21. K. H. W. J. ten Tusscher, D. Noble, P. J. Noble, and A. V. Panfilov, "A model for human ventricular tissue," *Am J Physiol Heart. Circ Physiol*, vol. 286, pp. 1573–1589, 2004.

22. G. Seemann, D. L. Weiss, F. B. Sachse, and O. Dössel, "Electrophysiology and tension development in a transmural heterogeneous model of the visible female left ventricle," in *Functional Imaging and Modeling of the Heart*, vol. 3504/2005 of *Lecture Notes in Computer Science*, pp. 172–182, Springer Berlin / Heidelberg, 2005.

23. J. Heuser, "Diagram of a myocardial infarction." Wikipedia.org.

24. P. W. Macfarlane, D. Browne, B. Devine, E. Clark, E. Miller, J. Seyal, and D. Hampton, "Modification of acc/esc criteria for acute myocardial infarction," *Journal of Electrocardiology*, vol. 37, no. Suppl, pp. 98–103, 2004.

25. G. S. Wagner, P. Macfarlane, H. Wellens, M. Josephson, A. Gorgels, D. M. Mirvis, O. Pahlm, B. Surawicz, P. Kligfield, R. Childers, and L. S. Gettes, "Aha/accf/hrs recommendations for the standardization and interpretation of the electrocardiogram: Part vi: Acute ischemia/infarction a scientific statement from the american heart association electrocardiography and arrhythmias committee, council on clinical cardiology; the american college of cardiology foundation; and the heart rhythm society endorsed by the international society for computerized electrocardiology," *Journal of the American College of Cardiology*, vol. 53, no. 11, pp. 1003–1011, 2009.

26. R. H. Christenson, C. P. deFilippi, and D. Kreutzer, "Biomarkers of ischemia in patients with known coronary artery disease: Do interleukin-6 and tissue factor measurements during dobutamine stress echocardiography give additional insight?," *Circulation*, vol. 112, no. 21, pp. 3215–3217, 2005.

27. F. S. Apple, A. H. B. Wu, J. Mair, J. Ravkilde, M. Panteghini, J. Tate, F. Pagani, R. H. Christenson, M. Mockel, O. Danne, and A. S. Jaffe, "Future biomarkers for detection of ischemia and risk stratification in acute coronary syndrome," *Clinical Chemistry*, vol. 51, no. 5, pp. 810–824, 2005.

28. J. Ferrero, B. Trenor, J. Saiz, F. Montilla, and V. Hernandez, "Electrical activity and reentry in acute regional ischemia: insights from simulations," in *Proceedings of the 25th Annual International Conference of the IEEE Engineering in Medicine and Biology Society*, vol. 1(1), pp. 17–20, 2003.

29. B. Rodriguez, N. Trayanova, and N. D, "Modeling cardiac ischemia," *Annals of the New York Academy of Sciences*, vol. 1080, pp. 395–414, 2006.

30. D. L. Weiss, M. Ifland, F. B. Sachse, G. Seemann, and O. Dössel, "Modeling of cardiac ischemia in human myocytes and tissue including spatiotemporal electrophysiological variations," *Biomedizinische Technik*, vol. 54, pp. 107–125, 2009.

31. B. Hopenfeld, J. G. Stinstra, and R. S. Macleod, "Mechanism for st depression associated with contiguous subendocardial ischemia," *Journal of Cardiovascular Electrophysiology*, vol. 15, no. 10, pp. 1200–1206, 2004.

32. M. C. MacLachlan, J. Sundnes, and G. T. Lines, "Simulation of st segment changes during subendocardial ischemia using a realistic 3-d cardiac geometry," *IEEE Transactions on Biomedical Engineering*, vol. 52, no. 5, pp. 799–807, 2005.

33. J. Väisänen, J. Requena-Carrión, F. Alonso-Atienza, J. Hyttinen, J. L. Rojo-Álvarez, and J. Malmivuo, "Contributions of the 12 segments of left ventricular myocardium to the body surface potentials," *Functional Imaging and Modeling of the Heart*, vol. 4466/2007, pp. 300–309, 2007.

34. T. S. Ruud, *Body surface ST shifts generated by ischemic heart disease: a simulation study*. PhD thesis, University of Oslo, 2009.

35. W. T. Miller and D. B. Geselowitz, "Simulation studies of the electrocardiogram. I. the normal heart," *Circulation Research*, vol. 43, no. 2, pp. 301–315, 1978.

36. L. Tung, *A Bidomain Model for Describing Ischemic Myocardial D-C Potentials*. PhD thesis, M.I.T., Cambridge, Mass., 1978.

37. O. Dössel, *Lecture Slides: Bielektrische Signale und Felder*, ch. 4: 2D und 3D Ausbreitung "Bidomain Model". Universität Karlsruhe (TH), Institut für Biomedizinische Technik, 2005.

38. R. Plonsey, *Bioelectric Phenomena*, ch. 5, pp. 202–275. McGraw-Hill Book, Inc., 1969.

39. D. B. Geselowitz, "On bioelectric potentials in an inhomogeneous volume conductor," *Biophysical Journal*, vol. 7, pp. 7–11, 1967.

40. N. Ida and J. P. A. Bastos, *Electromagnetics and Calculation of Fields*. Springer, New York; Berlin; Heidelberg, 2 ed., 1997.

41. K. H. Huebner and E. A. Thornton, *The Finite-Element Method for Engineers*. Wiley, New York, 2 ed., 1982.

42. G. H. Golub and C. F. V. Loan, *Matrix Computations*. The Johns Hopkins University Press, 1996.

43. R. Weiss, *Parameter Free Iterative Linear Solvers*. Berlin: Akademie, 1 ed., 1996.

44. J. R. Shewchuk, "An introduction to the conjugate gradient method without the agonizing pain," tech. rep., School of Computer Science, Carnegie Mellon University, 1994.

45. R. Barrett, M. Berry, T. F. Chan, J. Demmel, J. Donato, J. Dongarra, V. Eijkhout, R. Pozo, C. Romine, and H. V. der Vorst, *Templates for the Solution of Linear Systems: Building Blocks for Iterative Methods*. SIAM, Philadelphia, PA, 2 ed., 1994.

46. Y. Jiang, W. Hong, Y. Meng, D. Farina, and O. Dössel, "Solving the inverse problem of electrocardiography in a realistic environment." Submitted to Annals of Biomedical Engineering, 2010.

47. Y. Jiang, C. Qian, R. Hanna, D. Farina, and O. Dössel, "Determination of optimal electrode positions of a wearable ECG monitoring system for detection of acute myocardial infarction: A simulation study." Submitted to Medical and Biological Engineering and Computing, 2009.

48. C. R. Johnson, "Direct and inverse bioelectric field problems," in *Computational Science Education Project,* DOE, 1995.

49. P. Franzone, B. Taccardi, and C. Viganotti, "An approach to inverse calculation of epicardial potentials from body surface maps," *Advances in Cardiology,* vol. 21, pp. 50–54, 1978.

50. R. Barr and M. Spach, "Inverse calculation of qrs-t epicardial potentials from body surface potential distributions for normal and ectopic beats in the intact dog," *Circulation Research,* vol. 42, pp. 661–675, 1978.

51. G. Huiskamp and A. v. Oosterom, "The depolarization sequence of the human heart surface computed from measured body surface potentials," *IEEE Transactions on Biomedical Engineering,* vol. 35, p. 1047, 1988.

52. A. van Oosterom, "Cell models – macroscopic source descriptions," in *Comprehensive Electrocardiology: Theory and Practice in Health and Disease,* vol. 1, pp. 155–179, Pergamon Press, 1989.

53. B. Tilg, G. Fischer, R. Modre, F. Hanser, B. Messnarz, M. Schocke, C. Kremser, T. Berger, F. Hintringer, and F. X. Roithinger, "Model-based imaging of cardiac electrical excitation in humans," *IEEE Transactions on Medical Imaging,* vol. 21, pp. 1031–1039, September 2002.

54. O. Skipa, M. Nalbach, F. Sachse, C. Werner, and O. Dössel, "Transmembrane potential reconstruction in anisotropic heart model," *International Journal of Bioelectromagnetism,* vol. 4, pp. 17–18, 2002.

55. B. Messnarz, B. Tilg, R. Modre, G. Fischer, and F. Hanser, "A new spatiotemporal regularization approach for reconstruction of cardiac transmembrane potential patterns," *IEEE Transactions on Biomedical Engineering,* vol. 51, pp. 273–281, February 2004.

56. G. Huiskamp and F. Greensite, "A new method for myocardial activation imaging," *IEEE Transactions on Biomedical Engineering,* vol. 44, no. 6, pp. 433–446, 1997.

57. A. N. Tikhonov and V. Y. Arsenin, *Solutions of ill posed problems.* Wiley, New York, 1977.

58. R. D. Fierro, G. H. Golub, P. C. Hansen, and D. P. O'Leary, "Regularization by truncated total least squares," *SIAM Journal on Scientific Computing,* vol. 18, no. 4, pp. 1223–1241, 1997.

59. G. Shou, L. Xia, M. Jiang, Q. Wei, F. Liu, and S. Crozier, "Truncated total least squares: A new regularization method for the solution of ecg inverse problems," *Biomedical Engineering, IEEE Transactions on,* vol. 55, pp. 1327–1335, April 2008.

60. C. C. Paige and M. A. Saunders, "Lsqr: Sparse linear equations and least squares problems," *ACM Transactions on Mathematical Software (TOMS),* vol. 8, no. 2, pp. 195–209, 1982.

61. M. E. Kilmer and D. P. O'Leary, "Choosing regularization parameters in iterative methods for ill-posed problem," *SIAM J. on Matrix Analysis and Applications,* vol. 22, pp. 1204–1221, 2001.

62. D. Calvetti, B. Lewis, and L. Reichel, "GMRes, L-curves, and discrete ill-posed problems," *BIT Numerical Mathematics,* vol. 42, no. 1, pp. 44–65, 2002.

63. C. Ramanathan, P. Jia, R. Ghanem, C. D., and Y. Rudy, "Noninvasive electrocardiographic imaging (ECGI): Application of the generalized minimal residual (GMRes) method," *Annals of Biomedical Engineering,* vol. 31, no. 8, pp. 981–994, 2003.

64. M. Jacobsen, *Modular Regularization Algorithms.* PhD thesis, Informatics and Mathematical Modelling, Technical University of Denmark, DTU, Richard Petersens Plads, Building 321, DK-2800 Kgs. Lyngby, 2004. Supervised by Prof. Per Christian Hansen.

65. M. Jiang, L. Xia, and G. Shou, *Advanced Intelligent Computing Theories and Applications. With Aspects of Contemporary Intelligent Computing Techniques,* ch. Combining Regularization Frameworks for Solving the Electrocardiography Inverse Problem, pp. 1201–1219. Communications in Computer and Information Science, Springer Berlin Heidelberg, 2007.

66. M. Jiang, L. Xia, and G. Shou, *Noninvasive Electroardiographic Imaging: Application of Hybrid Methods for Solving the Electrocardiography Inverse Problem,* vol. 4466/2007 of *Functional Imaging and Modeling of the Heart,* pp. 269–279. Springer Berlin / Heidelberg, 2007.

67. M. Foster, "An application of the wiener-kolmogorov smoothing theory to matrix inversion," *Journal of the Society for Industrial and Applied Mathematics,* vol. 9, no. 3, pp. 387–392, 1961.

68. R. O. Martin, *Inverse electrocardiography: Epicardial potentials.* PhD thesis, Duke University, Durham, NC, 1970.

69. R. O. Martin, T. C. Pilkington, and M. N. Morrow, "Statistically constrained inverse electrocardiography," *IEEE Transactions on Biomedical Engineering,* vol. 22, no. 6, pp. 487–492, 1975.

70. D. D. Jackson, "The use of a priori data to resolve non-uniqueness in linear inversion," *Geophys. J. R. astr. Soc.,* vol. 57, pp. 137–157, 1979.

71. A. van Oosterom, "The spatial covariance used in computing the pericardial potential distribution," in *Computational Inverse Problem of E lectrocardiography* (P. R. Johnston, ed.), vol. Advances in Computational Biomedicine 3, pp. 1–50, Southampton; U.K: Computational Mechanics Publications, 2001. Chapter 1.

72. P. C. Hansen, "Analysis of discrete ill-posed problems by means of the L-curve," *SIAM Review,* vol. 34, no. 4, pp. 561–580, 1992.

73. P. C. Hansen, "The L-curve and its use in the numerical treatment of inverse problems," in *Computational Inverse Problems in Electrocardiography*, ch. 4, pp. 119–142, WIT Press, Advances in Computational Bioengineering, 2001.

74. G. H. Golub, M. Heath, and G. Wahba, "Generalized cross-validation as a method for choosing a good ridge parameter," *Technometrics*, vol. 21, no. 2, pp. 215–223, 1979.

75. Y. Zhang, A. Ghodrati, and D. H. Brooks, "An analytical comparison of three spatio-temporal regularization methods for dynamic linear inverse problems in a common statistical framework," *Inverse Problem*, vol. 21, pp. 357–382, 2005.

76. F. Greensite, "The temporal prior in bioelectromagnetic source imaging problems," *IEEE Transactions on Biomedical Engineering*, vol. 50, pp. 1152–1159, October 2003.

77. P. C. Hansen, *Rank-deficient and discrete ill-posed problems: numerical aspects of linear inversion.* Society for Industrial and Applied Mathematics, 1999.

78. Y. Jiang, W. Hong, D. Farina, and O. Dössel, "Solving the inverse problem of electrocardiography in a realistic environment using a spatio-temporal LSQR-Tikhonov hybrid regularization method," in *IFMBE Proceedings World Congress on Medical Physics and Biomedical Engineeringl Physics and Biomedical Engineering*, vol. 25 (2), pp. 817–820, 2009.

79. S. Usui and H. Araki, "Wigner distribution analysis of bspm for optimal sampling," *IEEE Engineering in Medicine and Biology*, vol. 9, no. 1, pp. 29–32, 1990.

80. O. Dössel, F. Schneider, and M. Müller, "Optimization of electrode positions for multichannel electrocardiography with respect to electrical imaging of the heart," in *Proceedings of the 20th Annual International Conference of the IEEE Engineering in Medicine and Biology Society*, vol. 20 (1), pp. 71–74, 1998.

81. C. Hintermüller, M. Seger, B. Pfeifer, G. Fischer, R. Modre, and B. Tilg, "Sensitivity- and effort-gain analysis: multilead ECG electrode array selection for activation time imaging," *IEEE Transactions on Biomedical Engineering*, vol. 53, no. 10, pp. 2055–2066, 2006.

82. T. Arentz, J. von Rosenthal, T. Blum, J. Stockinger, G. Bürkle, R. Weber, N. Jander, F. J. Neumann, and D. Kalusche, "Feasibility and safety of pulmonary vein isolation using a new mapping and navigation system in patients with refractory atrial fibrillation," *Circulation*, vol. 108, pp. 2484–2490, 2003.

83. D. Corrado, C. Basso, L. Leoni, B. Tokajuk, B. Bauce, G. Frigo, G. Tarantini, M. Napodano, P. Turrini, A. Ramondo, L. Daliento, A. Nava, G. Buja, S. Iliceto, and G. Thiene, "Three-dimensional electroanatomic voltage mapping increases accuracy of diagnosing arrhythmogenic right ventricular cardiomyopathy/dysplasia," *Circulation*, vol. 111, pp. 3042–3050, 2005.

84. L. Gepstein and S. J. Evans, "Electroanatomical mapping of the heart: basic concepts and implications for the treatment of cardiac arrhythmias.," *Pacing and clinical electrophysiology : PACE*, vol. 21, no. 6, pp. 1268–1278, 1998.

85. E. Hoffmann, P. Nimmermann, C. Reithmann, F. Elser, T. Remp, and G. Steinbeck, "New mapping technology for atrial tachycardias," *2000*, vol. 4, pp. 117–120, Journal of Interventional Cardiac Electrophysiology.

86. D. L. Packer, "Three-dimensional mapping in interventional electrophysiology: Techniques and technology," *Journal of Cardiovascular Electrophysiology*, vol. 16, pp. 1110–1116, October 2005.

87. S. A. Ben-Haim, D. Osadchy, I. Schuster, L. Gepstein, G. Hayam, and M. E. Josephson, "Nonfluoroscopic, in vivo navigation and mapping technology," *Nature Medicine*, vol. 2, pp. 1393–1395, December 1996.

88. L. Gepstein, G. Hayam, and S. A. Ben-Haim, "A novel method for nonfluoroscopic catheter-based electroanatomical mapping of the heart," *Circulation*, vol. 95, pp. 1611–1622, 1997.

89. H. Kottkamp, B. Hügl, B. Krauss, U. Wetzel, A. Fleck, G. Schuler, and G. Hindricks, "Electromagnetic versus fluoroscopic mapping of the inferior isthmus for ablation of typical atrial flutter: A prospective randomized study," *Circulation*, vol. 102, pp. 2082–2086, 2000.

90. T. R. Betts, P. R. Roberts, S. A. Allen, A. P. Salmon, B. R. Keeton, M. P. Haw, and J. M. Morgan, "Electrophysiological mapping and ablation of intra-atrial reentry tachycardia after fontan surgery with the use of a noncontact mapping system," *Circulation*, vol. 102, pp. 419–425, 2000.

91. A. Kadish, J. Hauck, B. Pederson, G. Beatty, and C. Gornick, "Mapping of atrial activation with a noncontact, multielectrode catheter in dogs," *Circulation*, vol. 99, pp. 1906–1913, 1999.

92. V. Markides and D. W. Davies, "New mapping technologies: An overview with a clinical perspective," *Journal of Interventional Cardiac Electrophysiology*, vol. 13, pp. 43–51, August 2005.

93. P. G. Novak, L. Macle, B. Thibault, and P. G. Guerra, "Enhanced left atrial mapping using digitally synchronized navx three-dimensional nonfluoroscopic mapping and high-resolution computed tomographic imaging for catheter ablation of atrial fibrillation," *Heart Rhythm*, vol. 1, no. 4, pp. 521–522, 2004.

94. R. J. Schilling, N. S. Peters, and D. W. Davies, "Simultaneous endocardial mapping in the human left ventricle using a noncontact catheter: Comparison of contact and reconstructed electrograms during sinus rhythm," *Circulation*, vol. 98, pp. 887–898, 1998.

95. F. H. M. Wittkampf, E. F. D. Wever, R. Derksen, A. A. M. Wilde, H. Ramanna, R. N. W. Hauer, and E. O. R. de Medina, "Localisa : New technique for real-time 3-dimensional localization of regular intracardiac electrodes," *Circulation*, vol. 99, pp. 1312–1317, 1999.

96. L. Gepstein, A. Goldin, J. Lessick, G. Hayam, S. Shpun, Y. Schwartz, G. Hakim, R. Shofty, A. Turgeman, D. Kirshenbaum, and S. A. Ben-Haim, "Electromechanical characterization of chronic myocardial infarction in the canine coronary occlusion model," *Circulation*, vol. 98, pp. 2055–2064, 1998.

97. L. Gepstein, G. Hayam, and S. A. Ben-Haim, "Activation-repolarization coupling in the normal swine endocardium," *Circulation*, vol. 96, pp. 4036–4043, 1997.

98. S. Shpun, L. Gepstein, G. Hayam, and S. A. Ben-Haim, "Guidance of radiofrequency endocardial ablation with real-time three-dimensional magnetic navigation system," *Circulation*, vol. 96, pp. 2016–2021, 1997.

99. J. L. R. M. Smeets, S. A. Ben-Haim, L.-M. Rodriguez, C. Timmermans, and H. J. J. Wellens, "New method for nonfluoroscopic endocardial mapping in humans: Accuracy assessment and first clinical results," *Circulation*, vol. 97, pp. 2426–2432, 1998.

100. Y. Jiang, D. Farina, M. Bar-Tal, and O. Dössel, "An impedance based catheter positioning system for cardiac mapping and navigation," *IEEE Transactions on Biomedical Engineering*, vol. 56, no. 8, pp. 1963–1970, 2009.

101. M. D. Cerqueira, N. J. Weissman, V. Dilsizian, A. K. Jacobs, S. Kaul, W. K. Laskey, D. J. Pennell, T. Ryan, and M. S. Verani, "Standardized myocardial segmentation and nomenclature for tomographic imaging of the heart," *Circulation*, vol. 105, pp. 539–542, 2002.

102. D. Keller, G. Seemann, D. L. Weiss, and O. Dössel, "Detailed anatomical modeling of human ventricles based on diffusion tensor mri," in *Proceedings of Biomedizinische Technik*, 2006.

103. Y. Jiang, R. Hanna, D. Farina, and O. Dössel, "Noninvasive imaging of myocardial infarction from body surface potentials by solving the inverse ecg problem with a spatio-temporal maximum *a posteriori* regularization method." Submitted to Medical and Biological Engineering and Computing, 2010.

104. G. Li and B. He, "Localization of the site of origin of cardiac activation by means of a heart-model-based electrocardiographic imaging approach," *IEEE Transactions on Biomedical Engineering*, vol. 48, pp. 600–609, June 2001.

105. D. Farina and O. Dössel, "Non-invasive model-based localization of ventricular ectopic centers from multichannel ecg," *International Journal of Applied Electromagnetics and Mechanics*, vol. 30, no. 3-4, pp. 289–297, 2009.

106. A. van Oosterom, "The use of the spatial covariance in computing pericardial potentials," *IEEE Transactions on Biomedical Engineering*, vol. 46, no. 7, pp. 778–787, 1999.

107. D. Farina, Y. Jiang, O. Skipa, O. Dössel, C. Kaltwasser, and W. Bauer, "The use of the simulation results as a priori information to solve the inverse problem of electrocardiography for a patient," in *Proceedings of Computers in Cardiology*, vol. 32, pp. 571–574, September 2005.

108. Y. Serinagaoglu, D. H. Brooks, and R. S. MacLeod, "Improved performance of bayesian solutions for inverse electrocardiography using multiple information sources," *IEEE Transactions on Biomedical Engineering*, vol. 53, no. 10, pp. 2024–2034, 2006.

109. Y. Jiang, D. Farina, and O. Dössel, "Identification of the origin of premature ventricular contraction by means of inverse electrocardiographic problem using MAP-based regularization." To be submitted.

110. Y. Jiang, C. Qian, R. Hanna, D. Farina, and O. Dössel, "Optimization of the electrode positions of multichannel ECG for the reconstruction of ischemic areas by solving the inverse electrocardiographic problem," *International Journal of Bioelectromagnetism (Cover Article)*, vol. 11, no. 1, pp. 27–37, 2009.

Danksagung

Zuerst möchte ich mich bei Herrn Prof. Dr. rer. nat. Olaf Dössel für die Möglichkeit zur Durchführung meiner Doktorarbeit sowie für die großartige Unterstützung und Anregung ganz herzlich bedanken.

Herrn Prof. Dr. Bjørn Fredrik Nielsen möchte ich ebenfalls für das Interesse an meiner Arbeit, für die übernahme des Korreferats und für die kompetente Betreuung während meines Forschungsaufenthalts am Simula Research Laboratory in Oslo herzlich danken.

Herrn Dr.-Ing. Dmytro Farina möchte ich danken für die mehrjährige Zusammenarbeit und stetige Unterstützung.

Herrn Walther Schulze möchte ich danken für die Korrektur der Dissertation und die wertvollen Diskussionen.

Mein Dank gilt Herrn Raghed Hanna, Frau Verena Reimund, Herrn Meng Ye, Herrn Cong Qian, Herrn Wei Hong and Herrn Walther Schulze für die Unterstützung durch Studien- oder Diplomarbeit.

Mein Dank gilt auch der Firma Biosense Webster, die meine Promotion finanziell unterstützte. Insbesondere danke ich Herrn Meir Bar-Tal für die professionelle Zusammenarbeit und seine guten Ratschläge.

Die Förderung meines Auslandsaufenthalts in Oslo verdanke ich dem Karlsruhe House of Young Scientists (KHYS).

Nicht zuletzt möchte ich mich bei meiner Frau Yao Liu, meinen Eltern, meiner Schwester und meinem Sohn Ju-Lian für ihre unendliche Geduld, Nachsicht und Unterstützung bedanken.

Publication List: Journal Articles

1. Y. Jiang, D. Farina, M. Bar-Tal, and O. Dössel, "An impedance based catheter positioning system for cardiac mapping and navigation," *IEEE Transactions on Biomedical Engineering*, vol. 56, no. 8, pp. 1963–1970, 2009.
2. Y. Jiang, C. Qian, R. Hanna, D. Farina, and O. Dössel, "Optimization of the electrode positions of multichannel ECG for the reconstruction of ischemic areas by solving the inverse electrocardiographic problem," *International Journal of Bioelectromagnetism (Cover Article)*, vol. 11, no. 1, pp. 27–37, 2009.
3. D. Farina, Y. Jiang, and O. Dössel, "Acceleration of FEM-based transfer matrix computation for forward and inverse problems of electrocardiography," *Medical and Biological Engineering and Computing*, vol. 47, no. 12, pp. 1229–1236, 2009.

Publication List: Conference Contributions

1. Y. Jiang, D. Farina, C. Kaltwasser, O. Dössel, and W. R. Bauer, "Modeling and reconstruction of myocardial infarction," in *Proceedings of Biomedizinische Technik*, 2006.
2. Y. Jiang, D. Farina, and O. Dössel, "An improved spatio-temporal maximum a posteriori approach to solve the inverse problem of electrocardiography," in *Proceedings of Biomedizinische Technik*, 2007.
3. Y. Jiang, D. Farina, and O. Dössel, "Reconstruction of myocardial infarction using the improved spatio-temporal map-based regularization," in *Proceedings of the 6th International Symposium on Noninvasive Functional Source Imaging of the Brain and Heart and the International Conference on Functional Biomedical Imaging*, pp. 308–311, October 2007.
4. Y. Jiang, D. Farina, and O. Dössel, "Localization of the origin of ventricular premature beats by reconstruction of electrical sources using spatio-temporal map-based regularization," in *Proceedings of the 4th European Conference of the International Federation for Medical and Biological Engineering*, vol. 22, pp. 2511–2514, 2008.
5. Y. Jiang, D. Farina, and O. Dössel, "Effect of heart motion on the solutions of forward and inverse electrocardiographic problem - a simulation study," in *Proceedings of Computers in Cardiology*, vol. 35, pp. 365–368, 2008.
6. Y. Jiang, C. Qian, R. Hanna, D. Farina, and O. Dössel, "Optimization of the electrode positions of multichannel ECG for the reconstruction of ischemic areas by solving the inverse electrocardiographic problem," in *the 7th International Symposium on Noninvasive Functional Source Imaging of the Brain and Heart and the International Conference on Functional Biomedical Imaging*, 2009.
7. Y. Jiang, W. Hong, D. Farina, and O. Dössel, "Solving the inverse problem of electrocardiography in a realistic environment using a spatio-temporal LSQR-Tikhonov hybrid regularization method," in *IFMBE Proceedings World Congress on Medical Physics and Biomedical Engineering*, vol. 25, pp. 817–820, 2009.
8. Y. Jiang, C. Qian, R. Hanna, D. Farina, and O. Dössel, "Optimization of electrode positions of a wearable ecg monitoring system for efficient and effective detection of acute myocardial infarction," in *Proceedings of Computers in Cardiology*, vol. 36, pp. 293–296, 2009.
9. Y. Jiang, Y. Meng, D. Farina, and O. Dössel, "Effect of respiration on the solutions of forward and inverse electrocardiographic problems - a simulation study," in *Proceedings of Computers in Cardiology (Rosanna Degani Young Investigator Award Nomination)*, vol. 36, pp. 17–20, 2009.
10. Y. Jiang, D. Farina, and O. Dössel, "The inverse problem of electrocardiography in realistic environment," in *Conference on Applied Inverse Problems*, 2009.
11. D. Farina, Y. Jiang, O. Skipa, O. Dössel, C. Kaltwasser, and W. Bauer, "The use of the simulation results as a priori information to solve the inverse problem of electrocardiography for a patient," in *Proceedings of Computers in Cardiology*, vol. 32, pp. 571–574, September 2005.
12. D. Farina, Y. Jiang, and O. Dössel, "Noninvasive estimation of action potential duration dispersion within the ventricular tissue," in *Proceedings of Biomedizinische Technik*, 2007.
13. D. Farina, Y. Jiang, O. Dössel, C. Kaltwasser, and W. R. Bauer, "Model-based method of non-invasive reconstruction of ectopic focus locations in the left ventricle," in *Proceedings of the 4th European Congress for Medical and Biomedical Engineering*, vol. 22, pp. 2560–2563, 2008.
14. V. Reimund, D. Farina, Y. Jiang, and O. Dössel, "Reconstruction of ectopic foci using the critical point theory," in *Proceedings of the 4th European Congress for Medical and Biomedical Engineering*, vol. 22, pp. 2703–2706, 2008.
15. R. Hanna, Y. Jiang, D. Farina, and O. Dössel, "Imaging of cardiac electrical sources using a novel spatio-temporal map-based regularization method," in *IFMBE Proceedings World Congress on Medical Physics and Biomedical Engineering*, vol. 25, pp. 813–816, 2009.

16. W. Schulze, D. Farina, Y. Jiang, and O. Dössel, "A kalman filter with integrated tikhonov-regularization to solve the inverse problem of electrocardiography," in *IFMBE Proceedings World Congress on Medical Physics and Biomedical Engineering*, vol. 25, pp. 821–824, 2009.